LABORATORY LIFE

LABORATORY LIFE

The Construction of Scientific Facts

Bruno Latour · Steve Woolgar

Introduction by Jonas Salk

With a new postscript and index by the authors

PRINCETON UNIVERSITY PRESS
PRINCETON, NEW JERSEY

Published by Princeton University Press, 41 William Street, Princeton, New Jersey 08540

In the United Kingdom: Princeton University Press, Oxford

Copyright © 1979 by Sage Publications, Inc.
Copyright © 1986 by Princeton University Press

Library of Congress Cataloging in Publication Data will be found on the last printed page of this book

First Princeton Paperback printing, 1986
LCC 85-43378
ISBN 0-691-09418-7
ISBN 0-691-02832-X (pbk.)

Clothbound editions of Princeton University Press books are printed on acid-free paper, and binding materials are chosen for strength and durability. Paperbacks, while satisfactory for personal collections, are not usually suitable for library rebinding.

Printed in the United States of America by Princeton University Press, Princeton, New Jersey

9 8 7 6 5 4 3

CONTENTS

PREFACE TO SECOND EDITION

The most substantial change to the first edition is the addition of an extended postscript in which we set out some of the reactions to the book's first publication in the light of developments in the social study of science since 1979. The postscript also explains the omission of the term "social" from this edition's new subtitle. Other minor additions include a detailed Table of Contents, Additional References, and an Index. Readers tempted to conclude that the main body of the text replicates that of the original are advised to consult Borges (1981).

Wolvercote, August 1985

To The Salk Institute

*"If sociology could not be applied in a thorough going
way to scientific knowledge, it would mean that science
could not scientifically know itself."*
                                        —Bloor (1976)

"Méfiez-vous de la pureté, c'est le vitriol de l'âme."
                              —M. Tournier (Vendredi)

ACKNOWLEDGEMENTS

The field research which forms the basis for the discussion in this volume was carried out by the first author. A Fulbright Fellowship (1975-1976), a NATO Fellowship (1976-1977), and a special grant from the Salk Institute financed the field research. Special thanks are due to Professor Roger Guillemin and his group, who made the field work possible. Subsequent writing was financially assisted by PAREX, the Maison des Sciences de l'Homme and by Brunel University. It is a pleasure to acknowledge all those sources and to thank those who have taken the trouble to read sections of the work and make helpful criticisms.

INTRODUCTION

Scientists often have an aversion to what nonscientists say about science. Scientific criticism by nonscientists is not practiced in the same way as literary criticism by those who are not novelists or poets. The closest one comes to scientific criticism is through journalists who have had an education in science, or through scientists who have written about their own personal experiences. Social studies of science and philosophy of science tend to be abstract or to deal with well-known historical events or remote examples that bear no relationship to what occurs daily at the laboratory bench or in the interactions between scientists in the pursuit of their goals. In addition, journalistic or sociological accounts seem sometimes to have the sole purpose of proving merely that scientists are also human.

A love-hate relationship exists toward scientists in some segments of society. This is evident in accounts that deal with facets ranging from tremendously high expectations of scientific studies to their cost and their dangers—all of which ignore the content and process of scientific work itself. In the name of "science policy," studies of scientific activity by economists and sociologists are often concerned with numbers of publications and with duplication of effort. While such examinations are of some value, they leave much to be desired because, in part, the statistical tools are crude and these exercises are often aimed at controlling productivity and creativity. Most importantly, they are not concerned with the substance of scientific thought and scientific work. For these reasons, scientists are not drawn to read what outsiders have to say about science and much prefer the views of scientists about scientific endeavors.

However, the present book is somewhat different from accounts usually written by nonscientists about science. It's based on a two-

year study by a young French philosopher which was carried out at
The Salk Institute for Biological Studies and which was subsequently
written up in collaboration with an English sociologist. Although I
was not responsible for the initial invitation, I welcomed the opportunity to see if the approach that was contemplated would remedy
some of the shortcomings of previous social studies of science.

The approach chosen by Bruno Latour was to become part of a
laboratory, to follow closely the daily and intimate processes of
scientific work, while at the same time to remain an "inside" outside
observer, a kind of anthropological probe to study a scientific
"culture"—to follow in every detail what the scientists do and how and
what they think. He has cast what he observed into his own concepts
and terms, which are essentially foreign to scientists. He has translated
the bits of information into his own program and into the code of this
profession. He has tried to observe scientists with the same cold and
unblinking eye with which cells, or hormones, or chemical reactions
are studied—a process which may evoke an uneasy feeling on the part
of scientists who are unaccustomed to having themselves analyzed
from such a vantage point.

The book is free of the kind of gossip, innuendo, and embarrassing
stories, and of the psychologizing often seen in other studies or
commentaries. In this book the authors demonstrate what they call the
"social construction" of science by the use of honest and valid
examples of laboratory science. This in itself is an achievement for
they are, in a sense, laymen to laboratory science and are not expected
to grasp its fundamentals, but merely expected to comprehend only
that which is easiest to understand, such as the superficial aspects of
laboratory life.

In reading this book about my colleagues who have been observed
under a sociologist's microscope, I realized how "scientific" a study of
science could be when viewed by an outsider who felt impelled to
imitate the scientific approach he observed. The authors' tools and
concepts are crude and qualitative, but their will to understand
scientific work is consistent with the scientific ethos. Their courage,
and even brashness, in this undertaking reminds me of many scientific
endeavors in which nothing stands in the way of the pursuit of an
inquiry. This kind of objective observation by an outsider of scientists
at work, as if they were a colony of ants or of rats in a maze, could be
unbearable. However, this seems not to be so, and for me the most
interesting part of the work and of its outcome, is that Bruno Latour, a
philosopher-sociologist, began a sociological study of biology and

along the way came to see sociology *biologically*. His own style of thought was transformed by our concepts and ways of thinking about organisms, order, information, mutations, etc. Curiously, instead of sociologists studying biologists, who in turn are studying life processes —in a sort of infinite regression—here are sociologists coming to recognize that their work is only a subset of our own kind of scientific activity, which in turn is only a subset of life in the process of organization.

The final point, intended to suggest that this book is not unworthy of the attention of scientists, is in the bridge made between science or scientists and the rest of society. The word "bridge" is not quite right and I doubt that it would be acceptable to the authors because they claim to go much further. One of their main points is that the social world cannot exist on one side and the scientific world on the other because the scientific realm is merely the end result of many other operations that are in the social realm. "Human affairs" are not different from what the authors call "scientific production," and the chief accomplishment they claim is to reveal the way in which "human aspects" are excluded from the final stages of "fact production." I have doubts about this way of thinking and, in my own work, find many details which do not fit this picture, but I am always stimulated by attempts to show that the two "cultures" are, in fact, only one.

Whatever objection may be raised about the details and by the author's arguments, I am now convinced that this kind of direct examination of scientists at work should be extended and should be encouraged by scientists themselves in our own best interest, and in the best interest of society. Science, in general, generates too much hope and too much fear, and the history of the relationship of scientists and nonscientists is fraught with passions, sudden bursts of enthusiasm, and equally sudden fits of panic. If the public could be helped to understand how scientific knowledge is generated and could understand that it is comprehensible and no more extraordinary than any other field of endeavor, they would not expect more of scientists than they are capable of delivering, nor would they fear scientists as much as they do. This would clarify not only the social position of scientists in society, but also the public understanding of the substance of science, of scientific pursuits and of the creation of scientific knowledge. It is sometimes discouraging that although we dedicate our lives to the extension of knowledge, to shedding light and exemplifying rationality in the world, the work of individual scientists, or the work of scientists in general, is often understood only in a sort of magical or mystical way.

Even if we do not agree with the details of this book, or if we find it slightly uncomfortable or even painful in places, the present work seems to me to be a step in the right direction toward dissipating the mystery that is believed to surround our activity. I feel certain that in the future many institutes and laboratories may well include a kind of in-house philosopher or sociologist. For myself, it was interesting to have Bruno Latour in our institute, which allowed him to carry out the first investigation of this kind of which I am aware and, most interestingly, to have observed the way in which he, and his approach, was transformed by the experience. It would be very useful for this critique itself to be criticized. This would both help the authors (and other scholars with similar interests and background) to assist scientists to understand themselves through the mirror provided, and help a wider public understand the scientific pursuit from a new and different and rather refreshing point of view.

—Jonas Salk, M.D.

La Jolla, California
February 1979

Chapter 1

FROM ORDER TO DISORDER

5 mins. John enters and goes into his office. He says something very quickly about having made a bad mistake. He had sent the review of a paper. . . . The rest of the sentence is inaudible.

5 mins. 30 secs. Barbara enters. She asks Spencer what kind of solvent to put on the column. Spencer answers from his office. Barbara leaves and goes to the bench.

5 mins. 35 secs. Jane comes in and asks Spencer: "When you prepare for I.V. with morphine, is it in saline or in water?" Spencer, apparently writing at his desk, answers from his office. Jane leaves.

6 mins. 15 secs. Wilson enters and looks into a number of offices, trying to gather people together for a staff meeting. He receives vague promises. "It's a question of four thousand bucks which has to be resolved in the next two minutes, at most." He leaves for the lobby.

6 mins. 20 secs. Bill comes from the chemistry section and gives Spencer a thin vial: "Here are your two hundred micrograms, remember to put this code number on the book," and he points to the label. He leaves the room.

Long silence. The library is empty. Some write in their offices, some work by windows in the brighly lit bench space. The staccato noise of typewriting can be heard from the lobby.

9 mins. Julius comes in eating an apple and perusing a copy of *Nature.*

9 mins. 10 secs. Julie comes in from the chemistry section, sits down on the table, unfolds the computer sheets she was carrying, and begins to fill in a sheet of paper. Spencer emerges from his office, looks over her shoulder and says: "hmm, looks nice." He then disappears into John's office with a few pages of draft.

9 mins. 20 secs. A secretary comes in from the lobby and places a newly typed draft on John's desk. She and John briefly exchange remarks about deadlines.

9 mins. 30 secs. Immediately following her, Rose, the inventory assistant, arrives to tell John that a device he wants to buy will cost three hundred dollars. They talk in John's office and laugh. She leaves.

Silence again.

10 mins. John screams from his office: "Hey Spencer, do you know of any clinical group reporting production of SS in tumour cells?" Spencer yells back from his office: "I read that in the abstracts of the Asilomar conference, it was presented as a well-known fact." John: "What was the evidence for that?" Spencer: "Well, they got an increase in . . . and concluded it was due to SS. Maybe, I'm not sure they directly tested biological activities, I'm not sure." John: "Why don't you try it on next Monday's bioassay?"

10 mins. 55 secs. Bill and Mary come in suddenly. They are at the end of a discussion. "I don't believe this paper," says Bill. "No, it's so badly written. You see, it must have been written by an M.D." They look at Spencer and laugh . . . (excerpt from observer's notes).

Every morning, workers walk into the laboratory carrying their lunches in brown paper bags. Technicians immediately begin preparing assays, setting up surgical tables and weighing chemicals. They harvest data from counters which have been working overnight. Secretaries sit at typewriters and begin recorrecting manuscripts which are inevitably late for their publication deadlines. The staff, some of whom have arrived earlier, enter the office area one by one and briefly exchange information on what is to be done during the day. After a while they leave for their benches. Caretakers and other workers deliver shipments of animals, fresh chemicals and piles of mail. The total work effort is said to be guided by an invisible field, or more particularly, by a puzzle, the nature of which has already been decided upon and which may be solved today. Both the buildings in which these people work and their careers are safeguarded by the Institute. Thus, cheques of taxpayers' money arrive periodically, by courtesy of the N.I.H., to pay bills and salaries. Future lectures and meetings are at the forefront of people's minds. Every ten minutes or so, there is a telephone call for one of the staff from a colleague, an editor, or some official. There are conversations, discussions, and arguments at the benches: "Why don't you try that?" Diagrams are scribbled on blackboards. Large numbers of computers spill out masses of print-out. Lengthy data sheets accumulate on desks next to copies of articles scribbled on by colleagues.

By the end of the day, mail has been dispatched together with manuscripts, preprints, and samples of rare and expensive substances packed in dry ice. Technicians leave. The atmosphere becomes more relaxed and nobody runs anymore. There are jokes in the lobby. One thousand dollars has been spent today. A few slides, like Chinese idiograms, have been added to the stockpile; one character has been deciphered, a miniscule, invisible increment. Minute hints have dawned. One or two statements have seen their credibility increase (or decrease) a few points, rather like the daily Dow Jones Industrial Average. Perhaps most of today's experiments were bungled, or are leading their proponents up a blind alley. Perhaps a few ideas have become knotted together more tightly.

A Philippino cleaner wipes the floor and empties the trash cans. It has been a normal working day. Now the place is empty, except for the lone figure of an observer. He silently ponders what he has seen with a mild sense of bewilderment . . . (Observer's Story).

Since the turn of the century, scores of men and women have penetrated deep forests, lived in hostile climates, and weathered hostility, boredom, and disease in order to gather the remnants of so-called primitive societies. By contrast to the frequency of these anthropological excursions, relatively few attempts have been made to penetrate the intimacy of life among tribes which are much nearer at hand. This is perhaps surprising in view of the reception and importance attached to their product in modern civilised societies: we refer, of course, to tribes of scientists and to their production of science. Whereas we now have fairly detailed knowledge of the myths and circumcision rituals of exotic tribes, we remain relatively ignorant of the details of equivalent activity among tribes of scientists, whose work is commonly heralded as having startling or, at least, extremely significant effects on our civilisation.

It is true, of course, that in recent years a wide variety of scholars have turned their attention to science. Frequently, however, their interest has focused on the large-scale effects of science. There are now a number of studies of the size and general form of overall scientific growth (e.g., Price, 1963; 1975), the economics of its funding (Mansfield, 1968; Korach, 1964), the politics of its support and influence (Gilpin and Wright, 1964; Price, 1954; Blisset, 1972), and the distribution of scientific research throughout the world (Frame et al., 1977). But it is easy to be left with the impression that research with such macroconcerns has enhanced rather than reduced the mystery of science. Although our knowledge of the external effects and reception of science has increased, our understanding of the complex activities which constitute the internal workings of scientific activity remains undeveloped. The emphasis on the external workings of science has been exacerbated by the application of concepts to science which are peculiar to social scientists of differing persuasions and theoretical commitments. Rather than making scientific activity more understandable, social scientists have tended through their use of highly specialised concepts to portray science as a world apart. A plethora of different specialised approaches have variously been brought to bear on science, such that the resulting overall picture is largely incoherent. Analyses of citations in scientific papers tend to tell us little about the substance of the papers; macroanalyses of science funding remain virtually silent on the nature of intellectual activity; quantitative histories of scientific development have tended to overemphasise those characteristics of science which most readily lend themselves to quantification. In addition, many of these approaches have too often accepted the products of science and taken them for granted in their subsequent analysis, rather than attempting to account for their initial production.

Our dissatisfaction with these approaches was considerably worsened by the realisation that very few studies of science have undertaken any kind of self-appraisal of the methods employed. This is surprising in that one might automatically expect students of science to be constantly aware of the basis for their pretensions to produce "scientific" findings: it might be reasonable to expect scholars concerned with the production of science to have begun to examine the basis for their own production of findings. Yet the best works of these scholars remain mute on their own methods and conditions of production. It can, of course, be argued that a lack of reflexivity is inevitable in an area which is still comparatively young, and that excessive attention to methodological issues would detract from the production of badly needed, albeit preliminary, research findings. But, in fact, the little evidence available suggests that new research areas do not usually postpone discussions of methodological issues in favour of the early production of substantive results. Rather, methodological clarification and discussion take place at an early stage of development (Mulkay et al., 1975). Perhaps a more plausible explanation of the lack of methodological reflexivity in social studies of science is simply that such an approach would be inconsistent with the dominance of macroconcerns noted already. Attention to the details of one's own methodology would thus constitute an enterprise radically different from concerns with overall development, or the implications of growth for science policy and funding.

Partly as a result of our dissatisfaction, and in an effort both to penetrate the mystique of science and to provide a reflexive understanding of the detailed activities of working scientists, we decided to construct an account based on the experiences of close daily contact with laboratory scientists over a period of two years (see Materials and Methods below).

The Observer and the Scientist

When an outside observer first expresses interest in the activities of working scientists, he can expect one of a variety of different reactions. If he is a fellow professional scientist working in a different field, or if he is a student working towards final admission into the scientific profession, the outsider will usually find that his interest is easily accommodated. Barring any circumstances involving extreme secrecy or competition between the parties, scientists can react to expressions of interests by adopting a teaching role. Outsiders can thus be told the basic principles of scientific work in a field which is relatively strange to them. However, for outsiders who are completely ignorant of science and do not aspire to join the ranks of professional scientists, the situation is rather different. The most naive (and perhaps least common) reaction is that nonscientific outsiders simply have no business probing the activities of science. More commonly, although working scientists realise that a variety of nonscientific outsiders, such as historians, philosophers, and sociologists can and do have professional interests in science, the precise point of their questions and observations is a source of some bewilderment. This is understandable in that working scientists do not normally possess more than outline knowledge of the principles, theories, methods, and issues at stake within disciplines other than their own. An observer who declares himself to be an "anthropologist of science" must be a source of particular consternation.

On the one hand, lack of knowledge can lead to marked disinterest in the reports produced by outsiders about science. A common response of this kind is that scholarly tracts in social studies of science seem "rather dull." If nothing else, this kind of comment provides a salient reminder of the perceived irrelevance for scientists of many social studies of science. On the other hand, lack of familiarity with disciplines outside natural science can provoke suspicion. Thus, it is often assumed that outsiders' interests must focus on the seedier aspects of scientific life because investigators are seen to be posing questions which are essentially irrelevant to practical scientific activity. Consequently, the fodder deemed most appropriate for such investigators tends to be tales of scandal and intrigue, of behaviour which fails the usual high standards of scientific enquiry or which is unethical, of the exchange of great ideas over coffee, or of renowned acts of genius and various eureka experiences. This is not to suggest that outsiders necessarily take such information at face value. Never-

theless, it is clear that the kind of information provided by scientists
will have a significant effect in shaping investigators' reports and that
the information provided depends, in turn, on the nature of the
relationship between scientist and investigator. It is important, therefore, to look briefly at the nature of this relationship and at the way it
may affect the production of reports about science.

We were fortunate that the discussion in this volume is informed by
research carried out at an institution with an avowedly well-developed
tradition for the cultivation of a wide range of scientific and philosophical interests. In particular, the founders had established the principle
that the institution should house research interests which encompassed a range of "life sciences" well beyond those of mainstream
biology. For example, a department of linguistics was conceived as an
integral part of the institution. Partly as a result of this general
principle, problems of initial access were considerably lessened.
Under the auspices of the head of one particular laboratory, one of us
was given office space for two years in immediate proximity to the day
to day activities of working scientists. However, despite the alleviation
of institutional obstacles to entry, the outside observer remained a
source of some puzzlement for members of the laboratory. What
exactly were his specific motives and objectives in studying the
laboratory?

It is perhaps tempting for an outside observer to present his interests
in terms of established categories of scholarly investigation, rather
than in a way which might exacerbate participants' curiosity or sense
of suspicion. For example, the label of "historian" or "philosopher"
might be more readily acceptable than either "sociologist" or "anthropologist." The term "anthropologist" is readily associated with the
study of "primitive" or "prescientific" belief systems. The term
"sociologist" gives rise to a plethora of different interpretations, but
essentially it can be seen by the working scientist to concern a range of
phenomena, all of which impinge in some way on matters of social and
political intrigue. Not surprisingly, therefore, the application of the
term "sociology" to a study of scientific activity will be regarded by
many scientists as dealing primarily with all these "nonscientific"
aspects of science. Sociological interest in science thus appears to
concern a variety of behavioural phenomena which fall into a residual
category: these phenomena unavoidably impinge upon scientific
practice by virtue of the fact that scientists are social beings; but they
are essentially peripheral to the practice itself. In this view, social

phenomena occasionally make their presence felt in instances of extreme secrecy, fraud, or on other relatively infrequent occasions. It is only then that the kernel of scientific logic and procedure is severely threatened and scientists find their work disrupted by the intrusion of external factors.

The Social and the Scientific: A Participant's Resource

A number of sources testify to the prevalence of this conception of sociology and "things social" among scientists. Firstly, this view is consistent with the relatively frequent perception by scientists that sociologists are engaged in some kind of scholarly muckraking. In response to enquiries from investigators who have declared their lack of scientific expertise, information is provided which concerns events essentially external to science. Secondly, a method commonly used by scientists to fault or cast doubt on the claims of others is to draw attention to the social circumstances of the production of the claim. For example, the assertion that

X observed the first optical pulsar

can be severely undermined by use of the following formulation:

X thought he had seen the first optical pulsar, having stayed awake three nights in a row and being in a state of extreme exhaustion.

In the second version, the inner logic of systematic scientific procedure has been disrupted by the intrusion of social factors. As we shall see in more detail in due course, "social factors" here refer both to "staying awake three nights" as well as to the transformation of a straight-forward "observation" into emphasis on the process of "thinking about seeing something." For the observation to have been successful, science should have proceeded either in isolation from such "social factors" or, as is sometimes the case with "great" scientists, in spite of them. Given the presence of such "social factors," no ordinary scientist can pursue science successfully. Observations, claims, and achievements can thus be explained away or faulted by the invocation of social circumstances. Thirdly, although the invocation of social circumstances can be used to detract from scientific achievement, it is also possible to recast social factors as an integral part of routine scientific procedure. As a result, the "social factors" in question no longer appear extraneous to science. Because they are no longer about the "social," these factors pass beyond the realm of sociological expertise. For example, in the case of the discovery of pulsars (Woolgar, 1978), a number of radio astronomy groups complained that their Cambridge rivals had unduly delayed the release of news of their discovery. In other words, attempts were made to lessen the nature of the Cambridge achievement by drawing attention to the way in which communication about the discovery had been handled. One of very many commentators made the following double-edged comment:

> The truth is that Hewish and the whole Cambridge group had for several months achieved a screen of security and secrecy which, in itself, was almost as much of an accomplishment as the discovery itself (Lovell, 1973:122).

By way of reply to similar criticisms, Cambridge spokesmen claimed that the need for secrecy was merely part of a normal scientific process:

> In the long history of science, it has, I think, been regarded as the right of an individual or group making a scientific discovery to follow up this discovery without any obligation to publish their first preliminary result (Ryle, 1975).

The argument here is that what had been regarded as grounds for casting doubt on the scientificity of Cambridge's conduct, was in fact integral to the normal process of science. Behaviour dubbed "secretive" (the term itself was hotly contested by Cambridge participants) was held to be a normal part of scientific procedure rather than an extraneous social factor which could be used to fault Cambridge behaviour. Moreover, several participants argued that because such behaviour was a normal part of the scientific process, it did not merit any special attention by sociological outsiders.

We shall return in due course to a detailed discussion of the use by scientists of similar procedures in dealing with the circumstances associated with their activities. But our argument is not just that the distinction between "social" and "intellectual" is prevalent among working scientists. More importantly, this distinction provides a resource upon which scientists can draw when characterising either their own endeavours or those of others. It is therefore important to investigate the nature of this distinction and the way it is used by scientists. The extent to which the distinction between "social" and "intellectual" is accepted as unproblematic by observers of science may have significant consequences for the reports about science which they produce.

The Social and the Scientific: The Observer's Dilemma

At one extreme, we can envisage the wholesale adoption by an observer of the distinction mentioned above. In this case, the observer holds an assumption that scientific phenomena occupy a realm largely distinct from that of social phenomena, and that it is only to the latter that the concepts, procedures, and expertise of sociology can be applied. As a result, the procedures and achievements central to scientists' work become largely immune from sociological explanation. Approaches which implicitly adopt this standpoint have been roundly criticised on several grounds. Rather than repeat these criticisms in detail, we shall merely outline some of the main critical themes. Firstly, the decision to concentrate only on "social" rather than "technical" aspects of science severely limits the range of phenomena that can be selected as appropriate for study. Put simply, this means that there is no point in doing sociology of science unless one can clearly identify the presence of some politician breathing down the necks of working scientists. Where there is no such obvious interference by external agencies, it is argued, science can proceed without the need for sociological analysis. This argument hinges on a particularly limited notion of the occasional influence of socio-political factors; the substance of science proceeds unaffected if such factors are absent. Secondly, emphasis on "social" in contradistinction to "technical" can lead to the disproportionate selection of events for analysis which appear to exemplify "mistaken" or "wrong" science. As we shall show, an important feature of fact construction is the process whereby "social" factors disappear once a fact is established. Since scientists themselves preferentially retain (or resurrect) the existence of "social" factors where things scientific are thought to have gone wrong, the adoption of the same viewpoint by an observer will necessarily lead him to the analysis of the way social factors affect, or have given rise to, "wrong" beliefs. As Barnes (1974) has argued, however, there is at least a very real need for a symmetrical approach to the analysis of beliefs (cf., Bloor, 1976). Scientific achievements held to be correct should be just as amenable to sociological analysis as those thought to be wrong. Thirdly, emphasis on the "social" has led commentators to argue for some redress of an imbalance: not enough attention is thought to have been paid to the "technical." For example, Whitley has argued that sociological interest in science is in danger of turning into a sociology of scientists rather than a fully fledged sociology of science:

> a separation of the study of producers of certain cultured artifacts, that is of science, without reference to the form and substance of science itself is mistaken (Whitley, 1972:61).

A fourth source of criticism addresses analyses inspired by Merton's portrayal of the normative structure of science. Many of these analyses exemplify sociologists' separation of "social" from "technical." Much criticism concerns the lack of empirical basis for the ethos of modern science which these analyses outline. It has, for example, been cogently argued that Merton's norms simply do not govern the behaviour of scientists in the way he suggests (Mulkay, 1969). More recently, it has been pointed out that the existence of both norms and counternorms in science (Mitroff, 1974) derives from the insufficiently critical appraisal by sociologists of scientists' statements to outsiders about their work (Mulkay, 1976). More important than this criticism of the empirical basis for scientists' norms, however, is the point that such sociological analyses ignore the technical substance of science. Even if the norms he specified were found to be correct, the sociologist might as well be describing a community of expert fishermen, for all he tells us about the nature or substance of their activity.

In an effort to pay more attention to the "technical" rather than the "social," Mulkay (1969) argues that the body of established knowledge and the associated "cognitive and technical norms" are a more realistic constraint on scientists' behaviour than are social norms. Consequently (Mulkay, 1972), scientists are known to be working within a system largely consistent with Kuhn's (1970) description of paradigm-bound research. The argument that "technical" factors merit treatment in the same fashion and to the same extent as do "social" factors has led to research which emphasises the investigation of *parallels* between social and intellectual development. It is thus axiomatic to several contributions in this area that an examination of cognitive developments should proceed in conjunction with an understanding of "concomitant" social developments. Perhaps the most obvious example of this formulation is in the work of Mullins (1972; 1973a; 1973b). Here social processes (for example, the emergence of "social organisation leaders") are seen to occur in tandem with developments on the "intellectual side" (for example, a shift between "defining a position" and "doing studies"). The discussion of social processes is presented quite separately from the treatment of intellectual developments. In a similar way, models of scientific growth have frequently presented areas of science as passing through various stages of development, each of which has attendant social and cognitive characteristics (Crane, 1972; Mulkay et al., 1975). The emphasis here is on producing "an account which shows some of the connections between intellectual development and social processes" (Mulkay et al., 1975: 188).

The investigation of scientific activity in terms of the connections between two different aspects of this activity leads to several difficulties. As already mentioned, some sociologists have complained because the correct balance between the "social" and the "intellectual" has not been achieved. For example, Law (1973) argues that Mullins (1972) concentrates less on the development of ideas than on changing network characteristics of a specialty over time (see also Gilbert, 1976: 200). At the same time, it is partly by virtue of the distinction between social and intellectual factors that the problem of causal relationship has arisen: does the formation of social groupings give rise to the pursuit by scientists of certain intellectual lines of enquiry, or does the existence of intellectual problems lead to the creation of social networks of scientists? Some authors avoid attempting to specify the direction of this causal relationship (Mulkay et al., 1975). Others have suggested that the direction varies according to the scientific area under investigation (for example, Edge and Mulkay, 1976: 382) and that it is a problem requiring further research (for example, Tobey, 1977: esp. footnote 4).

The commitment to an understanding of "technical" or "intellectual" issues provides an important challenge to traditional sociological research methods. This challenge has been taken up by Edge and Mulkay (1976), whose study of the emergence of radio astronomy in Britain provides a comprehensive history of detailed technical developments. As such, their report is a substantial departure from earlier sociology of science perspectives. It is interesting, however, that certain commentators have reviewed the report in terms of the relative emphasis given to the "social" and "technical" aspects of radio astronomy. Crane, for example, has said that the authors' emphasis on technical history has dwarfed the part of the argument devoted to theoretical interpretation and that there is a corresponding lack of adventure in the authors' attempts at generalisation:

> the authors present sociological analyses of some aspects of the development of the specialty but even here, as they themselves state, their discussion is "at a low level of generality and remains close to the empirical data generated in the case study" (Crane, 1977: 28).

For our purposes, an important aspect of the departure from earlier work stems from the cooperation in producing this report between an ex-member of a radio astronomy research group and a sociologist. Such cooperation would seem a sensible prerequisite for all outsiders' attempts to grapple seriously with the technical details of science. However, this cooperation is not without its own specific problems.

Mulkay (1974) argues that the sociological study of science requires a close examination of its technical culture and hence the active cooperation of technically competent participants. He also notes that because outsiders are seldom interested in technical culture and are usually technically incompetent, the accounts given them by participants must be treated with considerable caution. Scientists confronted by an audience of outsiders appear to convey a definite confusion in their accounts between *scientific* and *historical* accuracy. The relationship between scientist and outsider is highlighted by Mulkay's remarks on interviews variously conducted by ex-participant, sociologist and both together. Rapport can be quickly established between ex-participant and interviewer if the discussion concerns technical issues similar to those routinely discussed by the interviewee as part of his day-to-day activity. Discussion of more sociological issues was generally left until later in the interview and, especially where both ex-participant and sociologist were present, this exacerbated the interviewee's perception of the sociologist as an outsider. The interviewee assumed the sociologist to be qualified in areas of discussion which did not directly bear upon the technical content of his science.

These observations of difficulties experienced in the course of interaction with interviewees further support the idea that scientists themselves work with a very definite distinction between "social" and "technical." The same distinction can provide a problem for observers in that it raises the question of whether or not an equitable balance has been reached between the two sides of the dichotomy. This question remains, despite affirmations that "technical and social issues are intimately linked" (Mulkay, 1974: 114).

We should like to argue that it is not necessary to attach particular significance to the achievement of a "correct" balance between "social" and "intellectual" factors. This is for two main reasons. Firstly, as already mentioned, the distinction between "social" and "technical" factors is a resource drawn upon routinely by working scientists. Our intention is to understand how this distinction features in the activities of scientists, rather than to demonstrate that emphasis on one or the other side of the duality is more appropriate for our understanding of science. Secondly, our interest in the details of scientific activity cuts across the distinction between "social" and "technical" factors. We want to pay attention to "technical" issues in the sense that the use by scientists of "technical" and "intellectual" terminology is clearly an important feature of their activity. But we regard the use of such concepts as a phenomenon to be explained. More significantly, we view it as important that our explanation of scientific activity should not depend in any significant way on the uncritical use of the very concepts and terminology which feature as part of that activity.

The "Anthropology" of Science

The focus of our study is the routine work carried out in one particular laboratory. The majority of the material which informs our discussion was gathered from *in situ* monitoring of scientists' activity in one setting. Our contention is that many aspects of science described by sociologists depend on the routinely occurring minutiae of scientific activity. Historic events, breakthroughs and competition are examples of phenomena which occur over and above a continual stream of ongoing scientific activities. In Edge's (1976) terms, our most general objective is to shed light on the nature of "the soft underbelly of science": we therefore focus on the work done by a scientist located firmly at his laboratory bench.

In line with this perspective, a project took shape which we called, for want of a better term, an anthropology of science. We use this description to draw attention to several distinctive features of our approach.[1] Firstly, the term anthropology is intended to denote the preliminary presentation of accumulated empirical material. Without claiming to have given an exhaustive description of the activities of all like-minded practioners, we aim to provide a monograph of ethnographic investigation of one specific group of scientists. We envisaged a research procedure analogous with that of an intrepid explorer of the Ivory Coast, who, having studied the belief system or material production of "savage minds" by living with tribesmen, sharing their hardships and almost becoming one of them, eventually returns with a body of observations which he can present as a preliminary research report. Secondly, as has already been hinted, we attach particular importance to the collection and description of observations of scientific activity obtained in *a particular setting*. By our commitment to techniques of participant observation we hope to come to terms with a major problem which have thus far dogged understanding of science. Recently, there has been a growing dissatisfaction with outside observers' reliance on scientists' own statements about the nature of their work. Some participants have themselves argued that printed scientific communications systematically misrepresent the activity that gives rise to published reports (Medawar, 1964).[2] In a similar manner, Watkins (1964) complains that the "didactic dead-pan" style required of scientific reporting creates various difficulties in understanding how science is done. In particular, scientists who eschew the autobiographical form of reporting make it difficult for readers to appreciate the programme or context which provide the backdrop to reported work. Sociologists have noted that similar tendencies cause particular problems for the sociological understanding of historical context (Mulkay, 1974; Woolgar, 1976a; Wynne, 1976), although it is usually held that contradictory interpretations are reconciliable through sociological explanation (Mulkay, 1976; but see Woolgar, 1976b). These comments on the problems involved in the use of scientists' accounts find a parallel in discussions of the "craft" character of science. For example, Ravetz (1973) suggests that the nature of scientific activity is thoroughly misrepresented by the form of presentation which is used in the reporting of science. Not only do scientists' statements create problems for historical elucidation; they also systematically conceal the nature of the activity which typically gives rise to their research reports. In other words, the fact that scientists often change the manner and content of their statements when talking to outsiders causes problems both for outsiders' reconstruction of scientific events and for an appreciation of how science is done. It is therefore necessary to retrieve some of the craft character of scientific activity through in situ observations of scientific practice. More specifically, it is necessary to show through empirical investigation how such craft practices are organised into a systematic and tidied research report. In short, how is it that the realities of scientific practice become transformed into statements about how science has been done? We regard the prolonged immersion of an outside observer in the daily activities of scientists as one of the better ways in which this and similar questions can be answered. This also has the advantage that our descriptions of scientific activity have emerged as a result of the observer's experiences in the field. In other words, we have not chosen consciously to focus predominantly on any one of the technological, historical, or psychological aspects of what is observed. No attempt was made to delimit the area of competence prior to our discussion, and there was no prior hypothesis about a concept (or set of concepts) which might best explain what was to be encountered in the field. Thirdly, our use of "anthropology" denotes the importance of bracketing our familiarity with the object of our study. By this we mean that we regard it as instructive to apprehend as strange those aspects of scientific activity which are readily taken for granted. It is evident that the uncritical acceptance of the concepts and terminology used by some scientists has had the effect of enhancing rather than reducing the mystery which surrounds the doing of science. Paradoxically, our utilisation of the notion of anthropological strangeness is intended to dissolve rather than reaffirm the exoticism with which science is sometimes associated. This approach, together with our desire to avoid adopting the distinction between "technical" and "social," leads us to what might be regarded as a particularly irreverent approach to the analysis of science. We take the apparent superiority of the members of our laboratory in technical matters to be insignificant, in the sense that we do *not* regard prior cognition (or in the case of an ex-participant, prior socialisation) as a necessary prerequisite for understanding scientists' work. This is similar to an anthropologist's refusal to bow before the knowledge of a primitive sorcerer. For us, the dangers of "going native" outweigh the possible advantages of ease of access and rapid establishment of rapport with participants. Scientists in our laboratory constitute a tribe whose daily manipulation and production of objects is in danger of being misunderstood, if accorded the high status with which its outputs are sometimes greeted by the outside world. There are, as far as we know, no a priori reasons for supposing that scientists' practice is any more rational than that of outsiders. We shall therefore attempt to make the activities of the laboratory seem as strange as possible in order not to take too much for granted. Outsiders largely unfamiliar with technical issues may severely jeopardise their observational acumen by initially submitting themselves to an uncritical adoption of the technical culture.

Our particular use of an anthropological perspective on science also entails a degree of reflexivity not normally evident in many studies of science. By reflexivity we mean to refer to the realisation that observers of scientific activity are engaged in methods which are essentially similar to those of the practioners which they study. Of course, debates about whether and in what senses the social sciences can be scientific are the familiar stock-in-trade of many sociologists. Frequently, however, these debates have hinged on erroneous conceptions of the nature of scientific method culled from philosophers' partial accounts of the way science is practised. Although, for example, much has been made of whether social science can (or should) follow Popper or Kuhn, the correspondence of the descriptions of science provided by these authors to the realities of scientific practice is somewhat unclear, to say the least.[3] In our discussion, we shall sidestep these general issues and instead concentrate on specific problems which the scientific practitioner and the observer of scientific activity may have in common. This will entail making explicit, particularly in the latter part of the discussion, our awareness of certain methodological problems which we face in the construction and presentation of our discussion.

We have attempted to meet the above requirements of an anthropological perspective by basing our discussion on the experiences of an observer with some anthropological training but largely ignorant of science. By using this approach we hope to shed some light on the process of production within the laboratory and on the similarities with the approach of the observer.

It is unlikely that our discussion will tell working scientists anything they do not already know. We would not presume, for example, to reveal hitherto undiscovered facts about the details of scientific work to the subjects of our study. It is clear (as we show) that most members of our laboratory would admit to the kinds of craft activities which we portray. At the same time, however, our description of the way in which such craft activities become transformed into "statements about science" might constitute a new perspective on what working scientists know to be the case. We anticipate that hackles might rise where participants hold an obdurate commitment to descriptions of scientific activity formulated in terms of research reports. Often this commitment stems from the perceived utility of such statements in procuring funds or claiming other privileges. Objections will thus be forthcoming where our alternative version of the way science proceeds is seen potentially to undermine or threaten the securement of privileges. The investigation of the basis for beliefs or, as is a more accurate description of the present discussion, of the social construction of scientific knowledge, is frequently construed as an attempt to cast doubt on the beliefs or knowledge under study. Analysts often face this kind of mistaken perception in the sociological study of knowledge (for example, Coser and Rosenberg, 1964: 667). Our "irreverence" or "lack of respect" for science is not intended as an attack on scientific activity. It is simply that we maintain an agnostic position. We should emphasise, therefore, that we do not deny that science is a highly creative activity. It is just that the precise nature of this creativity is widely misunderstood. Our use of creative does not refer to the special abilities of certain individuals to obtain greater access to a body of previously unrevealed truths; rather it reflects our premise that scientific activity is just one social arena in which knowledge is constructed.

It might also be objected that the work of the particular laboratory we have studied is unusual in that it is relatively poor at the intellectual level; that its activity comprises routinely dull work, which is not typical of the drama and conjectural daring prevalent in other areas of scientific work. However, the Nobel Prize for Medicine was awarded to one of the members of our laboratory in 1977, soon after we began preparation of this manuscript. If the work of the laboratory is merely routine, then it is possible to receive what is perhaps the most prestigious kind of acclaim from the scientific community for the kind of routine work we portray.

It is perhaps relatively easy to show the intrusion of social factors in cases of borderline, controversial science, or where secrecy and competition are evident. This is because it is precisely in these situations that scientists can offer evidence of nonscientific or extra-technical interference with their work. As a result, it is tempting in these cases to explain the occurrence of the "technical" in terms of the "social." The work of our laboratory, however, constitutes "normal" science which is relatively free from obvious sociological events. We are less tempted, therefore, to try to tease out instances of gossip and scandal; no sociological muckraking is intended, nor do we claim that science devoid of such intrigue is unworthy of sociological attention.

So far we have discussed some ways in which our approach differs from many traditional sociological interests. In particular, we have adopted the notion of an anthropological study of science to denote the particular sense of our conception of the social. We are not concerned with a sociological analysis in the functionalist tradition which tries to specify norms governing scientists' behaviour. At the same time, we want to avoid a perspective which implicitly adopts a distinction between "social" and "technical" issues, however closely related these might be said to be. The use of such a distinction can be dangerous either because it fails critically to examine the substance of technical issues or because the effects of the social are only apparent in the more obvious instances of external disruption. More significantly, the use of this distinction fails to examine its importance as a resource for scientific activity. In addition, our collection of observations within the setting has led us to a kind of research primarily concerned with the details of scientific activity rather than with all-encompassing historical description. Our discussion concerns the social construction of scientific facts, with the proviso that we use "social" in a special sense which will become clear in the course of our argument. Obviously, we want to avoid the simplistic imposition of concepts in our attempts to make sense of our observations of science. For example, our concern with the "social" is not confined to those nontechnical observations amenable to the application of sociological concepts such as norms or competition. Instead, we regard the process of construction of sense implied by the application of sociological concepts as highly significant for our own approach. It is this process of construction of sense which forms the focus of our discussion. As a working definition, therefore, it could be said that we are concerned with the *social* construction of scientific knowledge in so far as this draws attention to the *process* by which scientists make sense of their observations.

Let us recap by using an example to illustrate what we mean by the process of making sense in the social construction of science. Sometime in late 1967, Jocelyn Bell, a research student at Cambridge radio astronomy laboratories, noted the persistent appearance of a strange section of "scruff" on the recorded output from apparatus designed to produce a sky survey of quasars. This statement is itself a highly condensed version of an account gleamed from a variety of sources, including discussions with Bell (Woolgar, 1976a). Sociologists of different persuasions and research styles would undoubtedly view this episode in a variety of different ways. Those primarily interested in norms, for example, might enquire how the communication of news of this finding was handled in the light of prevailing competitive pressures. To what extent did scientists live up to, or evade, norms of universality? Such an approach would leave intact the activity involved in Bell's perception. A more sophisticated approach might enquire as to the social circumstances prevalent at the time. What were the constraints in terms of availability of equipment which made Bell's observation appear remarkable? What were the characteristics of the organisation of radio astronomy at that stage of its development that gave Bell's observation a special significance? This approach would be more sophisticated in the sense that factors such as the organisation of research at Cambridge and participants' experience of past disputes would be examined for their influence on the observation and its subsequent interpretation. Given a different state of affairs, it could be argued, the observation would have been interpreted differently or might not have occurred at all.

In this particular example, it might be argued that if scrutiny of the recording had been automated or if Bell had been sufficiently socialised into realising that the persistent recurrence of scruff was impossible and hence nonnoticeable, the discovery of pulsars would have been much longer in coming. Technical events, such as Bell's observations, are thus much more than mere psychological operations; the very act of perception is constituted by prevalent social forces. Our interest, however, would be in the details of the observation process. In particular, we should like to know the method by which Bell made sense of a series of figures such that she could produce the account: "There was a recurrence of a bit of scruff." The processes which inform the initial perception can be dealt with psychologically. However, our interest would be with the use of socially available procedures for constructing an ordered account out of the apparent chaos of available perceptions.

The Construction of Order

Our interest in the way in which scientific order is constructed out of chaos arises from two main considerations. Firstly, from the fact that there are always available a number of alternative sociological features which might be invoked to explain the occurrence of a particular scientific action. Because any alternative can in principle be undermined or faulted, it may be preferable to change the focus so as to examine the way in which features are invoked so as to produce order. Secondly, outside observers appear to be in a position essentially similar to scientists in that they are also confronted with the task of constructing an ordered account out of a disordered array of observations. By capitalising on the reflexivity of the observer's situation, we hope to obtain an interesting analytical handle on our understanding of scientific practice. We shall thus argue that by realising and subsequently examining this essential similarity of method, the observer can better understand certain details of scientific activity. Let us elucidate each of these two points in turn.

The first point can be best demonstrated with the use of an example, again taken from the development of research on pulsars (Woolgar, 1978). The following utterance was used as part of the analysis of the reception and controversy stirred up by the initial discovery of pulsars:

> The discovery of the first pulsar was reported in February 1968 although the discovery itself seems to have been made in about a two-month period up to September 1967 (Hoyle, 1975).

On the one hand, this utterance can be used as evidence for the existence of a *complaint* that the Cambridge group had somehow violated scientific protocol by unduly delaying release of news of their discovery. The time lag between September 1967 and February 1968 is a "noticeable" (and hence noteworthy) feature as far as the author was concerned. It is perhaps noticeable either because the author feels piqued that members of another group did not make the discovery or because he feels the delay in reporting somehow hindered progress in investigating properties of pulsars. Alternatively, the same utterance could in principle be used as evidence for the *admiration* expressed by this author for the Cambridge group's ability to keep things under wraps for so long. The utterance may constitute admiration, again because the time period is a noteworthy or unusual feature. In this reading, however, the time period represents an achievement made against considerable odds; the fact that it was achieved facilitated the protection of a graduate student's first achievement and enhanced the progress of science unhindered by outside interference from the media or other observers.

In principle, the number of alternative readings of this particular utterance is very large. The number which will be accepted as plausible by an informed audience, however, will be constrained by the particular context which is brought to bear upon the reading of the utterance. In the same way, researchers with any knowledge of the particular research situation in question will (almost automatically) find that one of the two alternatives outlined above is the more plausible. It might, for example, be argued that the reading of the utterance as a *complaint* is more consistent with other available evidence than is it's reading as *admiration*. It could thus be said that Hoyle's comments were made in the aftermath of the award of the Nobel Prize for the pulsar discovery; that this resurrected Hoyle's dormant bitterness about his previous dealings with the Cambridge group; and that this is consistent with the interpretation of Hoyle making a complaint.[4] Of necessity, however, arguments for the consistency of one particular reading with other evidence depend, in some complex way, on readings of other utterances made by proponents of the argument. If asked to justify these "auxiliary" readings, proponents would be forced either to invoke yet further readings or to return to the original utterance for justification. In either case, requests for justification can never, in principle, be exhausted. In practice, of course, even persistent challengers yield their ground and a reading is produced. In other words, a particular reading is made for practical purposes at hand. The point here, however, is that *in principle* any alternative can be questioned. The fact that many observers would regard the reading of *complaint* as more plausible than that of *admiration* is largely irrelevant. Alternative readings are always possible and any one reading can always be undermined or faulted.

By extending this argument to the observer's use of any observation, rather than just an utterance, we can provide the following provisional formulation of a major theme of our discussion. The observer has to base his analysis on shifting ground. He is faced with the task of producing an ordered version of observations and utterances when each of his readings of observations and utterances can be counterbalanced with an alternative. In principle, then, the task of producing an incorrigible version of the actions and behaviour of the subjects of his study is hopeless. Nevertheless, we know that observers regularly produce such ordered versions for consumption by others. His production of order must therefore be done "for practical purposes," which means that he proceeds by evading or ignoring difficulties of principle.[5] If this is the case, then it becomes important to understand how observers routinely ignore the philosophical problem of the constant availability of alternative descriptions and readings. In other words, one reaction to the recognition of these fundamental problems is to investigate the methods and procedures by which observers produce ordered versions of the utterances and observations which they have accumulated. The focus of investigation from this point of view is the production of order.

It is not difficult to realise that the work of scientists may well involve similar problems of procedure. It became clear in the study of pulsar research, for example, that participants were divided over the correct interpretation of reports of the discovery made by one of the principal investigators at Cambridge (Woolgar, 1978). Some claimed that these reports demonstrated inconsistency and lack of clarity, which were evidence of willful concealment and secrecy; others denied that there was any inconsistency. Of course, the occurrence of practically achieved alternative readings is most obvious in situations of controversy. It is surely the case, nonetheless, that the unhesitant accomplishment of readings goes on throughout scientific activity. The elimination of alternative interpretations of scientific data and the rendering of these alternatives as less plausible is a central characteristic of scientific activity. Consequently, the practising scientist is likely to be as much involved with the task of producing ordered and plausible accounts out of a mass of disordered observations as is the outside observer. By paying more attention to the way in which we, as observers, produce the account you are now reading, we hope to gain an insight into some of the techniques used by scientists in their attempts to produce ordered accounts.

In sum, then, our discussion is informed by the conviction that a body of practices widely regarded by outsiders as well organised, logical, and coherent, in fact consists of a disordered array of observations with which scientists struggle to produce order. As we have mentioned, the adoption of the belief that science is well ordered has a corollary, that any study of its practice is relatively straightforward and that the content of science is beyond sociological study. However, we argue that both scientists and observers are routinely confronted by a seething mass of alternative interpretations. Despite participants' well-ordered reconstructions and rationalisations, actual scientific practice entails the confrontation and negotiation of utter confusion. The solution adopted by scientists is the imposition of various frameworks by which the extent of background noise can be reduced and against which an apparently coherent signal can be presented. The process whereby such frameworks are constructed and imposed is the subject of our study.

The above comments are intended to justify the emphasis in our discussion on the ways in which scientists produce order. This necessarily involves an examination of the methodical way in which observations and experiences are organised so that sense can be made of them. As already noted, we have every reason to believe that the accomplishment of this kind of task is no mean feat, as is clear from a consideration of the corresponding task faced by the observer when confronted by his field notes. The observer's task is to transform notes of the kind presented at the beginning of this chapter into an ordered account. But exactly how and where should the observer begin this transformation? It is clear that when seen through the eyes of a total newcomer, the daily comings and goings of the laboratory take on an alien quality. The observer initially encounters a mysterious and apparently unconnected sequence of events. In order to make sense of his observations, the observer normally adopts some kind of theme by which he hopes to be able to construct a pattern. If he can successfully use a theme to convince others of the existence of a pattern, he can be said, at least according to relatively weak criteria, to have "explained" his observations. Of course, the selection and adoption of "themes" is highly problematic. For example, the way in which the theme is selected can be held to bear upon the validity of his explanation; the observer's selection of a theme constitutes his method for which he is accountable. It is not enough simply to fabricate order out of an initially chaotic collection of observations; the observer needs to be able to demonstrate that this fabrication has been done correctly, or, in short, that his method is valid.

One of the many possible schemes designed to meet criteria of validity holds that descriptions of social phenomena should be deductively derived from theoretical systems and subsequently tested against empirical observations. In particular, it is important that testing be carried out in isolation from the circumstances in which the observations were gathered. On the other hand, it is argued that adequate descriptions can only result from an observer's prolonged acquaintance with behavioural phenomena. Descriptions are adequate, according to this perspective, in the sense that they emerge during the course of techniques such as participant observation. The descriptions thus produced, it is argued, are more likely to find some measure of congruence with the set of categories and concepts of participants under study. This latter version of adequate sociological method enjoys a number of variations, ranging from Glaser and Strauss' (1968) notion of "grounded theory" to the dictum of "phenomenologically oriented" sociology that investigators should "be true to the data" (see, for example, Tudor, 1976). The scheme which favours the deductive production of independently testable descriptions is oriented towards what has been called *etic* validation (Harris, 1968), that is, the audience who will ultimately assess the validity of a description is a community of fellow observers. The main advantage of this scheme is the comparative ease with which the reliability and replicability of descriptions can be assessed. By contrast, the scheme which favours the "emergence" of phenomeno- logically informed descriptions of social behaviour is most appropri- ately amenable to *emic* validation, that is, the ultimate decision about the adequacy of description rests with participants themselves. This has the advantage that descriptions produced by an observer are less likely to be mere impositions of categories and concepts which are alien to participants. At the same time, however, descriptions based on the categorical systems of participants in particular situations can provide problems for their generalisation to other situations. Further- more, the observer remains accountable to a community of fellow observers in the sense that they provide a check that he has correctly followed procedures for emic validation.

This simplistic distinction between methods for making sense of observations scarcely does justice to the range of methodological positions and debates current within sociology. Nevertheless, it helps clarify the diversity of approaches which can be adopted in the study of science. Very crudely, if Mertonian analyses depend on etic valida- tion, in that they pay relatively little attention to participants' technical culture, the approach exemplified by Edge and Mulkay more closely relies on emic validation, at least in the sense that participants would agree that these authors have correctly utilised their technical con- cepts and terminology. In general, observers reliant on emic validation will necessarily be concerned with whether or not they are correctly using the concepts employed by the subjects of their study. But overzealous concern with the correct usage of these concepts entails the danger of "going native": in the extreme case, an analysis of a tribe couched entirely in the concepts and language of the tribe would be both incomprehensible and unhelpful to all nonmembers of the tribe. Moreover, the dangers of going native are particularly marked in the study of science, both because, as analysts, we are inevitably caught up in social "science" traditions originating with explicit attempts to mimic the natural sciences and because of the currently widespread acceptance of the methods and achievements of science in the culture of which we are part. We also recognise the importance of taking seriously the concepts used by members of the laboratory. But as one way of resisting the temptation to go native, we shall attempt to explain participants' use of these concepts as a social phenomenon. In line with the principle of emic validation, then, our interest in the details of scientific activity, in the ways in which scientists produce order from disorder, leads us to an approach which relies on the emergence, from the circumstances of our study, of themes for discerning patterns in our observations. We attempt to capitalise on the experiences of observation of a laboratory *in situ*: by being close to localised scientific practices the observer has a preferential situation from which to understand how scientists themselves produce order. At the same time, we recognise that it is inappropriate merely to take for granted the concepts with which scientists work.

Materials and Methods

The materials on which the discussion in this book is based were obtained during field research carried out by the first author between October 1975 and August 1977. The choice of laboratory was determined mainly by the generosity of one of the senior members of the institute in providing office space, free access to most discussions and to all the archives, papers and other documents of the laboratory, and part-time employment as a technician in the laboratory. The twenty-one-month programme of participant observation yielded a large body of data, only a small fraction of which is used in the present discussion. In addition to the field notes (referred to throughout our discussion by the page and volume number of the field diary), an intensive analysis was made of all the literature produced by members of the laboratory. At the same time, a wide range of documents relevant to the daily activities of the laboratory was amassed: drafts of articles in preparation, letters between participants, memoranda, and various data sheets provided by participants. Formal interviews were also carried out with all members of the laboratory as well as with certain other scientists in the field working at other laboratories. These interviews supplemented the vast body of comments and information gleaned during informal discussions. The reflections of the observer, particularly on his work as a technician in the laboratory, provided a further source of data.

Preliminary analysis and writing began soon after initial participation. Partly because of the availability of office space within the laboratory, it was possible to carry on the work of writing without losing opportunities of taking part in discussions between participants or of observing other aspects of daily life in the laboratory.

No attempt was made to conceal the observational role. For example, it was made clear to participants that notes were being taken on all that went on in the laboratory. The observer discussed his preliminary drafts with participants and organised several seminar discussions in which visiting sociologists and philosophers of science interacted with members of the laboratory.[6]

In all but Chapter 3, which is historical in character (see below), names, dates, and places have been changed or replaced by initials so as to protect the anonymity of those involved. We also decided only to use those anecdotes and events that, in our judgment, were unlikely to cause social or political repercussions.

The Organisation of the Argument

It will be clear from the argument of this chapter that our very specific interest in laboratory life concerns the way in which the daily activities of working scientists lead to the construction of facts. Obviously, this particular interest differs substantially from existing perspectives on laboratories. Consequently, we shall not dwell on aspects such as the administrative organisation of laboratory work (Swatez, 1970), the influence of such organisation on creativity, the influence of laboratory organisation on scientists' careers (Lemaine and Matalon, 1969), nor on the nature of communication and patterns of information flow (Bitz et al., 1975).[7] Rather our interests focus on two major questions: How are the facts constructed in a laboratory, and how can a sociologist account for this construction? What, if any, are the differences between the construction of facts and the construction of accounts?

In Chapter 2, we portray the laboratory as seen through the eyes of a total newcomer. The notion of anthropological strangeness is used to depict the activities of the laboratory as those of a remote culture and to thus explore the way in which an ordered account of the laboratory life can be generated without recourse to the explanatory concepts of the inhabitants themselves. In order to emphasise the fictional nature of the account-generating process, we place the burden of this anthropological investigation on the shoulders of a fictional character: the visit to the laboratory is made by "the observer." Of course, the activities and interests of the laboratory can also be explained along a historical dimension. In particular, laboratory activity can be seen to hinge on what have been previously been constructed and accepted as facts. Against the backcloth of laboratory activity provided by our anthropological observer, therefore, Chapter 3 undertakes a close examination of the historical construction of one particular fact and of the implications for subsequent laboratory work. In Chapter 4, we move from an historical exposition of the construction of a fact to a consideration of the microprocesses of negotiation which take place continually in the laboratory. The construction of facts depends critically on these microprocesses, and yet the retrospective character-isation of scientific activity frequently replaces them with epis-temological descriptions of "thought processes" and "logical reason-ing." We therefore look closely at the relationship between these alternative portrayals of scientific activity and at the way in which one form of account becomes replaced by another. In Chapter 5, we turn our attention to the producers of facts. In particular, we look at the series of strategies taken up by members of the laboratory in their decisions to back the construction of one or other fact and in their efforts to enhance their ability further to invest in the construction of "new" facts.

By the end of Chapter 5, we are in a position to reconsider the laboratory as a system of fact construction. On the basis of the preceding discussion we then recap, in Chapter 6, the essential elements of the process whereby an ordered account is fabricated from disorder and chaos. Finally, we discuss the essential similarity between the construction of accounts which characterises the work of the laboratory and our own construction of an account which portrays the laboratory in this way.

NOTES

1. We make no attempt here systematically to relate our methodological procedures to those used in anthropological studies. For preliminary discussions on the relevance of anthropology for the study of science, see Horton (1967) and the readings in Wilson (1970). More recent discussions include Shapin (forthcoming) and Bloor (1978).

2. Medawar (1964) formulates his argument in terms of the "processes of thought" which are misrepresented through scientific reports. While agreeing with the general point that these reports are a source of considerable obfuscation, we have severe reservations about any quest for the "thought processes" which "underly" the construction of these reports. As we argue in detail in Chapter 4, explanations of scientific activity in terms of thought processes are themselves considerably misleading.

3. The point has been made by a number of authors. See, for example, the discussions in Lakatos and Musgrave (1970) and Bloor (1974; 1976).

4. This argument is developed at length in Woolgar (1978).

5. This theme is taken up again in Chapter 6 with reference to the game of "Go." At the beginning of the game, any move appears as possible, or as good, as any other.

6. The rationale for this strategy and its effects on the relationship between observer and participants will be discussed in detail elsewhere.

7. A number of French authors have recently discussed laboratory science. See, for example, Lemaine et al. (1977), Callon (1978). For a remarkable history of the biology laboratory in the eighteenth century, see Salomon-Bayet (1978).

Chapter 2

AN ANTHROPOLOGIST VISITS THE LABORATORY

When an anthropological observer enters the field, one of his most fundamental preconceptions is that he might eventually be able to make sense of the observations and notes which he records. This, after all, is one of the basic principles of scientific enquiry. No matter how confused or absurd the circumstances and activities of his tribe might appear, the ideal observer retains his faith that some kind of a systematic, ordered account is attainable. For a total newcomer to the laboratory, we can imagine that his first encounter with his subjects would severely jeopardise such faith. The ultimate objective of systematically ordering and reporting observations must seem particularly illusory in the face of the barrage of questions which first occur to him. What are these people doing? What are they talking about? What is the purpose of these partitions or these walls? Why is this room in semidarkness whereas this bench is brightly lit? Why is everybody whispering? What part is played by the animals who squeak incessantly in ante-rooms?

But for our partial familiarity with some aspects of scientific activity and our ability to draw upon a body of common sense assumptions, a flood of nonsensical impressions would follow the formulation of these questions. Perhaps these animals are being processed for eating. Maybe we are witnessing oracular prophecy through the inspection of rats entrails. Perhaps the individuals spending hours discussing scribbled notes and figures are lawyers. Are the heated debates in front of the blackboard part of some gambling contest? Perhaps the occupants of the laboratory are hunters of some kind, who, after patiently lying in wait by a spectograph for several hours, suddenly freeze like a gun dog fixed on a scent.

Such speculations and the questions which give rise to them appear nonsensical precisely because we as observers do presuppose some knowledge of what the laboratory could be doing. For example, it is possible to imagine the purpose of walls and partitions without ever having set foot in a laboratory. We attempt to make sense not by bracketing our familiarity with the setting but by using features which we perceive as common both to the setting and to our knowledge or previous experience. Indeed, it would be difficult to provide any sensible account of the laboratory without recourse to our taken-for-granted familiarity with some aspects of science.

Clearly, then, the observer's organisation of questions, observations, and notes is inevitably constrained by cultural affinities. Only a limited set of questions is relevant and hence sensible. In this sense, the notion of a *total* newcomer is unrealisable in practice. At another extreme, an observer's total reliance on scientists' versions of laboratory life would be unsatisfactory. A description of science cast entirely in terms used by scientists would be incomprehensible to outsiders. The adoption of scientific versions of science would teach us little that is new about science in the making; the observer would simply reiterate those accounts provided by scientists when they conduct guided tours of their laboratory for visitors.

In practice, observers steer a middle path between the two extreme roles of total newcomer (an unattainable ideal) and that of complete participant (who in going native is unable usefully to communicate to his community of fellow observers). This is not to deny, of course, that at different stages throughout his research he is severely tempted towards either extreme. His problem is to select a principle of organisation which will enable him to provide an account of the laboratory sufficiently distinct from those given by scientists themselves and yet of sufficient interest to both scientists and readers not familiar with biology. In short, the observer's principle of organisation should provide an Ariadne's thread in a labyrinth of seeming chaos and confusion.

In this chapter, we follow the trials and tribulations of a fictional character, "the observer,"[1] in his attempts to use the notion of literary inscription[2] as a principle for organizing his initial observations of the laboratory.

Literary Inscription

Although our observer shares the same broad cultural knowlege as scientists, he has never seen a laboratory before and has no knowledge of the particular field within which laboratory members are working. He is enough of an insider to understand the general purpose of walls, chairs, coats, and so on, but not enough to know what terms like TRF, Hemoglobin, and "buffer" mean. Even without knowledge of these terms, however, he can not fail to note the striking distinction between two areas of the laboratory. One area of the laboratory (section B on Figure 2.1) contains various items of apparatus, while the other (section A) contains only books, dictionaries, and papers. Whereas in section B individuals work with apparatus in a variety of ways: they can be seen to be cutting, sewing, mixing, shaking, screwing, marking, and so on; individuals in section A work with written materials: either reading, writing, or typing. Furthermore, although occupants of section A, who do not wear white coats, spend long periods of time with their white-coated colleagues in section B, the reverse is seldom the case. Individuals referred to as doctors read and write in offices in section A while other staff, known as technicians, spend most of their time handling equipment in section B.

Each of sections A and B can be further subdivided. Section B appears to comprise two quite separate wings: in the wing referred to by participants as the "physiology side" there are both animals and apparatus: in the "chemistry side" there are no animals. The people from one wing rarely go into the other. Section A can also be subdivided. On the one hand, there are people who write and engage in telephone conversations; on the other hand, there are those who type and dial telephone calls. This division, like the others, is marked by partitions. In one area (the library) eight offices surround the perimeter of a conference room with table, chairs, and a screen. In the other area ("the secretariat") there are typewriters and people controlling the flow of telephone calls and mail.

What is the relationship between section A ("my office," "the office," "the library") and section B ("the bench")? Consulting the

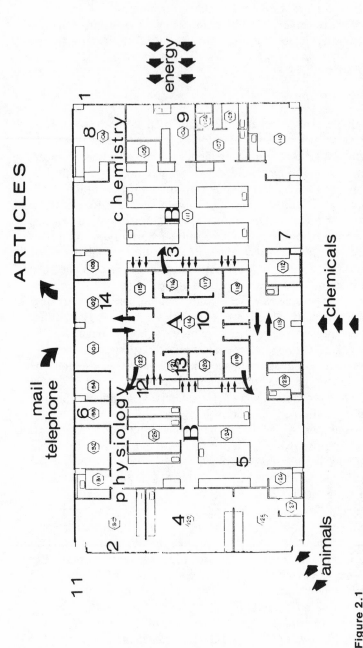

ARTICLES

Figure 2.1

Map of the laboratory showing partitions and the main flows described in the text. The numbers on the map correspond to photographs in the file (page 91). The map shows the extent to which the differences between section A and B, and between the chemistry and physiology wings, are reinforced by the architectural layout of the laboratory.

map he has drawn, our observer tries to imagine another institution or setting with a similar division. It is hard to call to mind any factory or administrative organisation which has a similar set up. If, for example, it was a factory, we might expect the office space (section A) to be much smaller. If it was some kind of administrative agency, the bench space (section B) would be entirely superfluous. Although the relation between the two wings of the office space is common to many productive units, the special relation between office space and bench space is sufficient to distinguish the laboratory from other productive units. This is apparent on two counts. Firstly, at the end of each day, technicians bring piles of documents from the bench space through to the office space. In a factory we might expect these to be reports of what has been processed and manufactured. For members of this laboratory, however, these documents constitute what is yet to be processed and manufactured. Secondly, secretaries post off papers from the laboratory at an average rate of one every ten days. However, far from being *reports* of what has been produced in the factory, members take these papers to be the *product* of their unusual factory. Surely, then, if this unit merely processes paper work, it must be some sort of administrative agency? Not so: even a cursory look at the papers shows that the figures and diagrams which they contain are the very same documents produced in section B a few days or weeks previously.

It occurs to our observer that he might be able to make sense of laboratory activity according to one very simple principle. For him, the scene shown in Photograph 13,[3] represents the prototype of scientific work in the laboratory: a desk belonging to one of the inhabitants of the office space (referred to as the doctors) is covered with paperwork. On the left is an opened issue of *Science.* To the right is a diagram which represents a tidied or summarised version of data sheets lying further to the right. *It is as if two types of literature are being juxtaposed:* one type is printed and published outside the laboratory; the other type comprises documents produced within the laboratory, such as hastily drawn diagrams and files containing pages of figures. Beneath the documents at the centre of the desk lies a draft. Just like the drafts of a novel or a report, this draft is scribbled, its pages heavy with corrections, question marks, and alterations. Unlike most novels however, the text of the draft is peppered with references, either to other papers, or to diagrams, tables or documents ("as shown in figure . . . ," "in table . . . we can see that . . . "). Closer inspection of the material lying on the desk (Photograph 13) reveals, for example, that the opened issue of *Science* is cited in the draft. Part of the argument contained in a *Science* article is said in the draft to be unrepeatable by virtue of what is contained in documents lying to the right of the desk. These documents are also cited in the draft. The desk thus appears to be the hub of our productive unit. For it is here that new drafts are constructed by the juxtaposition of two sources of literature, one originating outside and the other being generated within the laboratory.

It is no surprise to our observer to learn that scientists read published material. What surprises him more is that a vast body of literature emanates from within the laboratory. How is it that the costly apparatus, animals, chemicals, and activities of the bench space combine to produce a written document, and why are these documents so highly valued by participants?

After several further excursions into the bench space, it strikes our observer that its members are compulsive and almost manic writers. Every bench has a large leatherbound book in which members meticulously record what they have just done against a certain code number. This appears strange because our observer has only witnessed such diffidence in memory in the work of a few particularly scrupulous novelists. It seems that whenever technicians are not actually handling complicated pieces of apparatus, they are filling in blank sheets with long lists of figures; when they are not writing on pieces of paper, they spend considerable time writing numbers on the sides of hundreds of tubes, or pencilling large numbers on the fur of rats. Sometimes they use coloured papertape to mark beakers or to index different rows on the glossy surface of a surgical table. The result of this strange mania for inscription is the proliferation of files, documents, and dictionaries. Thus, in addition to the Oxford dictionary and the dictionary of known peptides, we can also find what might be called material dictionaries. For example, Photograph 2 shows a refrigerator which houses racks of samples, each of which bears a label with a ten-figure code number. Similarly, in another part of the laboratory, a vast supply of chemicals has been arranged in alphabetical order on shelves from which technicians can select and make use of appropriate substances. A more obvious example of these material dictionaries is the collection of preprints (Photograph 14, background) and thousands of files full of data sheets, each of which also has its own code number. Quite apart from these labelled and indexed collections is the kind of paperwork (such as invoices, pay cheques, inventory schedules, mail files, and so on) which can be found in most modern productive units.

When the observer moves from the bench space to the office space, he is greeted with yet more writing. Xeroxed copies of articles, with words underlined and exclamation marks in the margins, are everywhere. Drafts of articles in preparation intermingle with diagrams scribbled on scrap paper, letters from colleagues and reams of paper spewed out by the computer in the next room; pages cut from articles are glued to other pages; excerpts from draft paragraphs change hands between colleagues while more advanced drafts pass from office to office being altered constantly, retyped, recorrected, and eventually crushed into the format of this or that journal. When not writing, the occupants of section A scribble on blackboards (Photograph 10) or dictate letters, or prepare slides for their next talk.

Our anthropological observer is thus confronted with a strange tribe who spend the greatest part of their day coding, marking, altering, correcting, reading, and writing. What then is the significance of those activities which are apparently not related to the marking, writing, coding, and correcting? Photograph 4, for example, shows two young women handling some rats. Despite the protocol sheet to the right, the numbered tubes on the rack and the clock in the foreground which controls the rhythmn of the assay, the women themselves are neither writing nor reading. The woman on the left is injecting a liquid with a syringe and withdrawing another liquid with another syringe which she then passes on to the other woman; the second woman then empties the syringe into a tube. It is only then that writing takes over: the time and tube number is carefully recorded. In the meantime animals have been killed and various materials, such as ether, cotton, pipettes, syringes, and tubes have been used. What then is the point of killing these animals? How does the consumption of materials relate to the writing activity? Even the careful monitoring of the contents of the rack (Photograph 5) makes the situation no clearer to our observer. Over a period of several days, tubes are arranged in rows, other liquids are added, the mixtures are shaken and eventually removed for refrigeration.

Periodically, the routine of manipulation and rearrangement of tubes is interrupted. The samples extracted from rats are put into one of the pieces of apparatus and undergo a radical transformation: instead of modifying or labelling the samples, the machine produces a sheet of figures (Photograph 6). One of the participants tears the sheet from the machine's counter and, after scrutinising it carefully, arranges for the disposal of the tubes. In other words, the same tubes which had been carefully handled for a week, which had cost time and effort to the tune of several hundred dollars, were now regarded as worthless. The focus of attention shifted to a sheet of figures. Fortunately, our observer was quite used to finding such absurd and erratic behaviour in the subjects of his studies. Relatively unperturbed, therefore, he braced himself for his next surprise.

It was not long in coming. The sheet of figures, taken to be the end result of a long assay, was used as the input to a computer (Photograph 11). After a short time, the computer printed out a data sheet and it was this, rather than the original sheet of figures, which was regarded as the important product of the operation. The sheet of figures was merely filed alongside thousands like it in the library. Nor was the series of transformations yet complete. Photograph 12 shows a technician at work on several data sheets produced by the computer. Soon after this photograph was taken, she was called into one of the offices to show the product of her labours: a single elegant curve carefully drawn on graph paper. Once again, the focus of attention shifted: the computer data sheets were filed away and it was the peaks and slopes of the curve which excited comment from participants in their offices: "how striking," "a well differentiated peak," "it goes down quite fast," "this spot is not very different from this one." A few days later, the observer could see a neatly redrawn version of the same curve in a paper sent out for possible publication. If accepted, this same figure would be seen by others when they read the article and it was more than likely that the same figure would eventually sit on some other desk as part of a renewed process of literary juxtaposition and construction.

The whole series of transformations, between the rats from which samples are initially extracted and the curve which finally apears in publication, involves an enormous quantity of sophisticated apparatus (Photograph 8). By contrast with the expense and bulk of this apparatus, the end product is no more than a curve, a diagram, or a table of figures written on a frail sheet of paper. It is this document, however, which is scrutinised by participants for its "significance" and which is used as "evidence" in part of an argument or in an article. Thus, the main upshot of the prolonged series of transformations is a document which, as will become clear, is a crucial resource in the construction of a "substance." In some situations, this process is very much shorter. In the chemistry wing in particular, the use of certain pieces of apparatus makes it easy to get the impression that substances directly provide their own "signatures" (Photograph 9). While participants in the office space struggle with the writing of new drafts, the laboratory around them is itself a hive of writing activity. Sections of muscle, light beams, even shreds of blotting paper activate various recording equipment. And the scientists themselves base their own writing on the written output of the recording equipment.

It is clear, then, that particular significance can be attached to the operation of apparatus which provides some kind of written output. Of course, there are various items of apparatus in the laboratory which do not have this function. Such "machines" transform matter between one state and another. Photograph 3, for example, shows a rotary evaporator, a centrifuge, a shaker, and a grinder. By contrast, a number of other items of apparatus, which we shall call "inscription devices,"[4] transform pieces of matter into written documents. More exactly, an inscription device is any item of apparatus or particular configuration of such items which can transform a material substance into a figure or diagram which is directly usable by one of the members of the office space. As we shall see later, the particular arrangement of apparatus can have a vital significance for the production of a useful inscription. Furthermore, some of the components of such a configuration are of little consequence by themselves. For example, the counter shown in Photograph 6 is not itself an inscription device since its output is not directly usable in an argument. It does, however, form part of an inscription device known as a bioassay.[5]

An important consequence of this notion of inscription device is that inscriptions are regarded as having a direct relationship to "the original substance." The final diagram or curve thus provides the focus of discussion about properties of the substance. The intervening material activity and all aspects of what is often a prolonged and costly process are bracketed off in discussions about what the figure means. The process of writing articles about the substance thus takes the end diagram as a starting point. Within the office space, participants produce articles by comparing and contrasting such diagrams with other similar diagrams and with other articles in the published literature (see pp. 69-86).

At this point, the observer felt that the laboratory was by no means quite as confusing as he had first thought. It seemed that there might be an essential similarity between the inscription capabilities of apparatus, the manic passion for marking, coding, and filing, and the literary skills of writing, persuasion, and discussion. Thus, the observer could even make sense of such obscure activities as a technician grinding the brains of rats, by realising that the eventual end product of such activity might be a highly valued diagram. Even the most complicated jumble of figures might eventually end up as part of some argument between "doctors." For the observer, then, the laboratory began to take on the appearance of a system of literary inscription.

From this perspective, many hitherto strange occurrences fell into place. Many other types of activity, although superficially unrelated to the literary theme, could be seen as means of obtaining inscriptions. For example, the energy inputs (Photograph 1) represented inter-mediary resources to be consumed in the process of ensuring that inscription devices functioned properly. By also taking into account the supply of animals and chemicals, it was clear that a cycle of production which ended in a small folder of figures might have cost several thousand dollars. Similarly, the technicians and doctors who comprised the work force represented one further kind of input necessary for the efficient operation of the inscription devices and for the production and dispatch of articles.

The central prominence of documents in our discussion so far contrasts markedly with a tendency evident in some sociology of science to stress the importance of informal communication in scientific activity. For example, it has been frequently noted that the communication of scientific information occurs predominantly through informal rather than formal channels (Garvey and Griffith, 1967; 1971). This is particularly likely where there exists a well-developed network of contacts as, for example, in an invisible college (Price, 1963; Crane, 1969; 1972). Proponents of this argument have often played down the role of formal communication channels in informa-tion transfer, choosing instead to explain their continued existence in terms of an arena for the establishment of priority and subsequent conferral of credit (Hagstrom, 1965). Observations of the present laboratory, however, indicate that some care needs to be exercised in interpreting the relative importance of different communication chan-nels. We take formal communication to refer to highly structured and stylised reports epitomised by the published journal article. Almost without exception, every discussion and brief exchange observed in the laboratory centred around one or more items in the published literature (Latour, 1976). In other words, informal exchanges invariably focussed on the substance of formal communication. Later we shall suggest that much informal communication in fact establishes its legitimacy by referring or pointing to published literature.

Every presentation and discussion of results entailed the manipulation either of slides, protocol sheets, papers, preprints, labels, or articles. Even the most informal exchanges constantly focussed either directly or indirectly on documents. Participants also indicated that their telephone conversations nearly always focussed on the discussion of documents; either on a possible collaboration in the writing of a paper, or on a paper which had been sent but which contained some ambiguity, or on some technique presented at a recent meeting. When there was no direct reference to a paper, the purpose of the call was often to announce or push a result due to be included in a paper currently being prepared. Even when they were not discussing a draft, individuals devoted considerable energy to devising ways of attaining some readable trace. In these kinds of discussions, scientists anticipated that possible objections to their argument might appear in some forthcoming paper. More important for the present, however, is the omnipresence of literature in the sense that we have defined it, that is, in terms of written documents, only a few of which appear in published form.

The Culture of the Laboratory

To those familiar with the work of the laboratory, the above account will have little to say that is new. For an anthropologist, however, the notion of literary inscription is still problematic. As we said earlier, our observer has an intermediary status: while the broad cultural values which he shares with the scientists facilitate some familiarity with the commonplace objects and events in the laboratory, he is unwilling solely to rely on scientists' own versions of the way the laboratory operates. One consequence of his intermediary status is that his account so far has failed to satisfy any one audience. It could be said, for example, that in portraying scientists as readers and writers he has said nothing of the *substance* of their reading and writing. Indeed, our observer incurred the considerable anger of members of the laboratory, who resented their representation as participants in some literary activity. In the first place, this failed to distinguish them from any other writers. Secondly, they felt that the important point was that they were writing *about* something, and that this something was "neuroendocrinology." Our observer experienced the depressing sensation that his Ariane's thread had led him up a blind alley.

ARTICLES ABOUT NEUROENDOCRINOLOGY

We noted earlier that participants made sense of their juxtaposition of literatures by reference to a world of literature published outside the laboratory. To the extent that such literature represents the scriptures (Knorr, 1978) from which participants take the sense of their activities, we can only begin to understand what the literature is about by close inspection of the mythology which informs their activities. Our use of the term mythology is not intended pejoratively. Rather, it refers to a broad frame of reference within which can be situated the activities and practices of a particular culture (Barthes, 1957).

Our observer noticed that when asked by a total stranger, members of the laboratory replied that they worked (or were) "in neuroendocrinology." They went on to explain that neuroendocrinology was the result of a hybridisation which had taken place in the 1940s between neurology, described as the science of the nervous system, and endocrinology, the science of the hormonal system. It occurred to our observer that such location "in a field" facilitated the correspondence between a particular group, network, or laboratory and a complex mixture of beliefs, habits, systematised knowledge, exemplary achievements, experimental practices, oral traditions, and craft skills. Although referred to as the "culture" in anthropology, this latter set of attributes is commonly subsumed under the term paradigm when applied to people calling themselves scientists.[6] Neuroendocrinology seemed to have all the attributes of a mythology: it had had its precursors, its mythical founders, and its revolutions (Meites et al., 1975). In its simplest version, the mythology goes as follows: After World War II it was realised that nerve cells could also secrete hormones and that there is no nerve connection between brain and pituitary to bridge the gap between the central nervous system and the hormonal system. A competing perspective, designated the "hormonal feedback model" was roundly defeated after a long struggle by participants who are now regarded as veterans (Scharrer and Scharrer, 1963). As in many mythological versions of the scientific past, the struggle is now formulated in terms of a fight between abstract entities such as models and ideas. Consequently, present research appears based on one particular conceptual event, the explanation of which only merits scant elaboration by scientists. The following is a typical account: "In the 1950s there was a sudden crystallization of ideas, whereby a number of scattered and apparently unconnected results suddenly made sense and were intensely gathered and reviewed."

The mythology through which a culture represents itself is not necessarily entirely false. A count of publications, for example, shows that the growth of papers dealing with neuroendocrinology after 1950 was exponential, and that neuroendocrinology, which made up only 3 percent of endocrinology as a whole in 1968, grew to 6 percent by 1975. In broad outline, then, the growth of neuroendocrinology appears to have followed the pattern of what some sociologists of science have termed "scientific development" (for example, Crane, 1972; Mulkay et al., 1975). However, the mythology of its development is very rarely mentioned in the course of the day-to-day activities of members of the laboratory. The beliefs that are central to the mythology are noncontroversial and taken for granted, and only enjoy discussion during the brief guided tours of the laboratory provided for visiting laymen. In the setting, it is difficult to determine whether the mythology is never alluded to simply because it is a remote and unimportant remnant of the past or because it is now a well-known and generally accepted item of folklore.

After his first few days in the setting, our observer was no longer told about neuroendocrinology. Instead, daily concerns focussed on a different set of specific cultural values which, although from time to time talked about as being in neuroendocrinology, appeared to constitute a distinct culture (or "paradigm"). Our criteria for identifying this specific culture is not simply that a specialty represents a subset of a larger discipline. This would be no more accurate than considering the Bouarées' nations as a subset of the larger Boukara ethnical group. Instead, we use culture to refer to the set of arguments and beliefs to which there is a constant appeal in daily life and which is the object of all passions, fears, and respect. Participants in our laboratory said that they were dealing with "substances called releasing factors" (for popular accounts, see Guillemin and Burgus, 1972; Schally et al., 1973; Vale, 1976). When they presented their work to scientifically informed outsiders, they formulated their efforts as attempting "to isolate, characterize, synthesise and understand the modes of action of releasing factors." This is the brief that distinguishes them from their other colleagues in neuroendocrinology. It is also their cultural trait, their particularity, and their horizon of work and achievement. The general mythology provides them with the tenet that the brain controls the endocrine system, and they share this with a large cultural group of neuroendocrinologists. Specific to their own culture, however, is an additional postulate that "control by the brain is mediated by discrete chemical substances, so called releasing factors, which are of a peptidic nature" (Meites, 1970).[7] Their skills, working habits, and the apparatus at their disposal are all organised around one specific material (the hypothalamus), which is deemed especially important for the study of releasing factors.

Our observer can now picture his informants as readers and writers of Neuroendocrinological literature who acknowledge certain texts published in the previous five years as major achievements. These tests record the structure of several releasing factors in sentences comprising words or phonemes which relate to substances called amino acids. In general, the structure of any substance of a peptidic nature can be expressed in the form of a string of amino acids (for example, Tyr-Lys-Phe-Pro).[8] The texts that specify the structure of the first releasing factors were considered major breakthroughs by all informants (see Chapter 3). "In 1969 we discovered the structure of the thyrotropin releasing factor"; in 1971 they discovered or confirmed the structure of another releasing factor known as LRF; in 1972 they discovered the structure of a third substance called somatostatin (for general accounts, see Wade, 1978; Donovan et al., forthcoming).

The importance of articles specifying the structure of releasing factors is shown by the number of other articles which resulted. Papers written by other informants constituted the outside literature used in conjunction with internally produced inscriptions to generate new papers. Figure 2.2 shows the relative boom in the number of papers dealing with various substances after the initial specification of structure in so-called breakthrough papers. As a result of these publication explosions, the proportion of releasing factor publications in neuroendocrinology rose from 17 percent in 1968 to 38 percent in 1975. This suggests that the releasing factor "specialty" was responsible for the general increase in the importance of neuroendocrinology as a whole. Because of burgeoning outside interests, the laboratory's share of publication in the specialty actually decreased as a result of its success, from 42 percent in 1968 to 7 percent in 1975.[9] To put this in perspective, however, it is worth noting that in 1975 publications in releasing factors represented 39 percent of all publications in neuroendocrinology; neuroendocrinology represented only 6 percent of endocrinology as a whole, and endocrinology is only one of many disciplines within biology. Put another way, publications by members of the laboratory in 1975 represented only 0.045 percent of those in endocrinology. Clearly, some caution should be exercised in general-

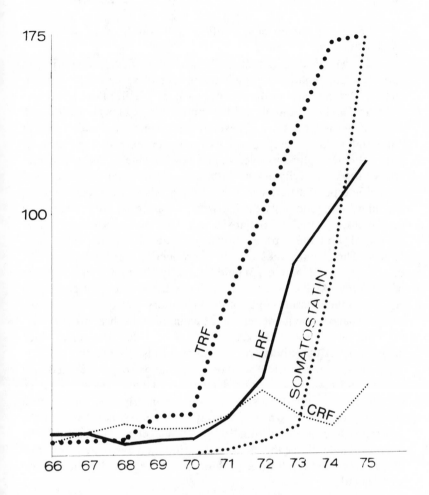

Figure 2.2
This diagram shows the number of articles published per year on each of four different releasing factors. The computation is based on the SCI, Permuterm and by combining the various spellings of releasing factors. The names chosen in this diagram are those used in the laboratory studied. TRF in 1970, LRF in 1971, and somatostatin in 1973, all show the same abrupt ascending curve. CRF, the structure of which is still unknown, is included for comparison.

ising the characteristics of scientific activity on the basis of this one laboratory.

So far we have said that each inscription device comprises a particular combination of machines, pieces of apparatus, and technicians. Articles are written on the basis of specific flow of outside literature, and with the use (either implicit or explicit) of part of the archives in the laboratory. These archives comprise a wide range of "material dictionaries," brain extracts, for example, as well as protocol books. Our observer should now be able to discern several distinct lines of activity in the laboratory, each of which corresponds to a specific type of article which is finally produced. For each type he should be able to identify the individuals concerned, their location in the laboratory, the technicians who assist, the inscription devices employed, and the type of outside literature to which their work relates. Three main lines of article production, referred to by participants as "programmes," could be clearly differentiated at the time of the study. As can be seen from Table 2.1, they do not contribute equally to the overall output of the laboratory, nor do they have the same cost and subsequent impact. By examining the three programmes in some detail, our observer hoped to be able to specify which characteristics of activity were peculiar to this laboratory.

The *first* type of article written in this laboratory concerned *new natural* substances in the hypothalamus (see Chapter 3). A substance is obtained by superimposing two sets of inscriptions, one from a recording device known as an assay in the physiology side of the laboratory and the other from "purification cycles" carried out in the chemistry side. Since the assay and purification cycle are inscription devices common to all three programmes, we shall describe them in some detail.

Despite the many different types of activity referred to as assays (for example, the bioassay, the *in vitro* and *in vivo* assays, direct or indirect assays, radioimmunological or biological assays) they are all based on the same principle (Rodgers, 1974). A recording mechanism (such as a myograph, a gamma counter, or a simple rating sheet) is connected up to an organism (either a cell, a muscle, or a whole animal) so as to produce an easily readable trace. A substance with a known effect on the organism is then administered to the organism as a control. The effect on the organism is inscribed and its recorded trace is taken as a baseline. An unknown substance is then administered and its effect recorded. The result is a recorded *difference* between two traces, a

Table 2.1

First Programme (isolation of new substance)	31 papers	15% of total	24 c.p.i.
Second Programme: Total (analogs and functions)	78 papers	37% of total	—
Task One (analogs)	—	—	—
Task Two (structure function)	52 papers	24% of total	7.6 c.p.i.
Task Three (clinical)	19 papers	9% of total	21 c.p.i.
Task Four (basic chemistry)	7 papers	3% of total	7.2 c.p.i.
Third Programme (mode of actions)	47 papers	22% of total	10.6 c.p.i.
Technical Papers	20 papers	9% of total	7 c.p.i.
General Articles	27 papers	13% of total	9 c.p.i.
Others	10 papers	5% of total	—
Total	213 papers		
Mean			12.4 c.p.i.

difference about which simple perceptive judgments ("it is the same," "it goes up," "there is a peak") can be made. If there is a difference, it is taken as the sign of an "activity" in the unknown substance. Since the central objective of the culture is to define any activity in terms of a discrete chemical entity, the unknown substance is taken to the other side of the laboratory for tests in the second main type of inscription device, the purification cycle.

The goal of the purification cycle is to isolate the entity which is believed to have caused the recorded difference between two traces. Samples of brain extract are subjected to a series of *discriminations* (Anonymous, 1974). This entails the use of some stationary material (such as a gel or a piece of blotting paper) as a selective sift which delays the gradual movement of a sample of brain extract. (This movement can be variously due to gravity, electric forces, or cellular binding—Heftmann, 1967.) As a result of this process, samples are transformed into a large number of fractions, each of which can be scrutinised for physical properties of interest. The results are recorded in the form of several peaks on graph paper. Each of these peaks represents a discriminated fraction, one of which may correspond to the discrete chemical entity which caused an activity in the assay. In order to discover whether the entity is present, the fractions are taken back to the physiology section of the laboratory and again take part in an assay. By superimposing the result of this last assay with the result of the previous purification, it is possible to see an overlap between one peak and another. If the overlap can be repeated, the chemical fraction is referred to as a "substance" and is given a name.

Ideally, this shuttle between the assay (Photograph 4) and the purification cycle (Photograph 7) ends with the identification of an "isolated" substance. This is almost never the case, however, because most of the differences between activities in the assay disappear when the assay is repeated. The postulated substance CRF, for example, has been shuttling to and fro in six laboratories since 1954 (cf., Figure 2.2). Even when differences between activities do not disappear, the entity can often no longer be traced after a few steps of purification. As we shall see later, the elimination of these elusive and transitory substances (known as "artefacts") is the main concern of the tribe. Although the details of the elimination process are extremely complex, the general principle is simple.

Since most competitors' claims to have an "isolated" substance are put into quotation marks, it follows that the assertion that an entity is "isolated" depends primarily on the operation of local criteria. When this claim has been made within the laboratory, the chemical fraction breaks out of the shuttle between assay and purification and switches to another circuit of operations. This new circuit comprises an inscription device known as an Amino Acid Analyser (AAA), which automatically records the effects of the isolated sample on a series of other chemical "reagents" and allows this effect to be directly read in terms of certain letters of the amino acid vocabulary. Thus, the inscription of the substance is decipherable in letters, such as, for example, Glu, Pyro, His, rather than just in terms of peaks, spots, and slopes. However, this is not the end of the matter. By this stage, each component amino acid is known; but the particular order of the amino acids has not yet been determined. To do this, the previous samples are taken to another room, where there are expensive inscription devices handled by full-time "PhD holders." The two main inscription devices, the "mass spectrometer" and the "Edmann degradation sequence," provide written spectra and diagrams which allow the specification of the configuration of amino acids which are present in the substance. These are great and rare moments in the work of the first programme. The determination of structure constitutes the most exciting and exhausting periods of work, which are remembered vividly by participants many years after. In the next chapter we shall follow the history of one of these substances in detail and return to a closer explanation of the activities mentioned here.

The concern of a *second* main programme in the laboratory is to reconstruct substances (whose structure has already been determined), using amino acids supplied by the chemical industry, and to evaluate their activity. The main objective of this programme is to produce artificially reconstructed substances, known as analogs, with properties which, because they are different from the original substances, will facilitate their use in medicine or physiology. The second research programme can be divided into four tasks.[10] The first task is the chemical production of analogs. Instead of buying analogs or obtaining them from another investigator, the laboratory can supply substances relatively cheaply in its own inhouse chemical section. The production of analogs is largely mechanised, using apparatus such as the peptide automatic synthetizer. Many of the analytical inscription devices (such as the mass spectrometer, the amino acid analyser, or the nuclear magnetic resonance spectrometer) which are used in the original purification of a substance are also used in its artificial reconstruction. In the second programme, however, these inscription devices are used to monitor the reconstruction process rather than to produce new information. The second task concerns so-called "structure function relationships." Using a number of slightly different analogs, physiologists try to identify connections between bioassay effects and combinations of analogs which give rise to them. For example, the natural substance which inhibits the release of a substance called growth hormone, is a fourteen amino acid structure. By substituting a right-handed form for the left-handed form of the amino acid at the eighth position, a more potent substance is obtained. This has major implications for the treatment of diabetes. Consequently, the outcome of these kinds of trial and error operations, which make up 24 percent of published papers, are of special interest to funding agencies and to the chemical industry (Latour and Rivier, 1977). A third task, which makes up to 9 percent of published papers, concerns the determination of structure function relationships in the effect of substances on humans. Most of the papers which result from this work are written in collaboration with clinicians. The aim is to devise analogs which most nearly match the natural substances required for clinical purposes. It would be desirable, for example, to devise an analog of LRF which would inhibit the release of LH instead of triggering it. This would make possible the production of a much better contraceptive pill than at present and thus represents a highly prized (and highly funded) research objective. The fourth and last task, which makes up only 3 percent of total research output, comprises research in collaboration with fundamental chemists on the configuration of molecules which make up the substance. The role of the laboratory in this work is mainly the provision of material, but the results are nevertheless very important for studies of "structure-function relationships."[11] As in the third task, first authors of papers resulting from this fourth task are based outside the laboratory.

So far we have discussed two main programmes: the isolation of new natural substances on the one hand and their reproduction by synthesis on the other. A *third* programme is said by participants to be aimed at understanding the mechanisms by which different substances interact. This work is carried out in the physiology section of the laboratory using bioassays. A variety of different trails, ranging from those generating crude behavioural responses to those which record the rate of DNA synthesis following hormonal contact, are used to try and assess how substances react together.

In terms of published papers, these three programmes accounted respectively for 15 percent, 37 percent, and 22 percent of the total output from the laboratory between 1970 and 1976. It is rarely the case, however, that participants refer to the programme in which they are working. The specification and particular arrangement of apparatus does not in itself correspond to the self-perceptions of work which they hold. Rather than saying, "I am doing purification," for example, they are much more likely to say, "I am purifying substance X." It is not purification in general which concerns them, but "the isolation of CRF"; it is not the synthesis of analogs, but the study of "D TRP 8 SS." Furthermore, objectives of each programme change in the course of a few months. Our notion of programme is thus inadequate in that it is merely an intermediary device which our observer has used in becoming familiar with his setting. On the other hand, our observer now knows what distinguishes this laboratory from others and which papers are written on the basis of particular combinations of staff and inscription devices. We reserve for later discussion an appraisal of laboratory activity in terms of specific individuals, careers, historical periods, and items of apparatus.

THE "PHENOMENOTECHNIQUE"

We have so far related how our observer apprehended the laboratory in terms of the prevalence of written documents and of inscription devices. In particular, the notion of literature provided an organising principle with which the observer could make sense of his observations without relying solely on participants' accounts. "Literature" refers both to the central importance accorded a variety of documents and to the use of equipment to produce inscriptions which are taken to be about a substance, and which are themselves used in the further generation of articles and papers. In order to explicate the notion of literary inscription as applied to apparatus, we shall provide an inventory of the material setting of the laboratory.

One important feature of the use of inscription devices in the laboratory is that once the end product, an inscription, is available, all the intermediary steps which made its production possible are forgotten. The diagram or sheet of figures becomes the focus of discussion between participants, and the material processes which gave rise to it are either forgotten or taken for granted as being merely technical matters.[12] A first consequence of the relegation of material processes to the realm of the merely technical is that inscriptions are seen as direct indicators of the substance under study. Especially in apparatus such as the amino acid analyser (Photograph 9), the substance appears to inscribe its own signature (Spackmann et al., 1958). A second consequence, however, is the tendency to think of the inscription in terms of confirmation, or evidence for or against, particular ideas, concepts, or theories.[13] There thus occurs a transformation of the simple end product of inscription into the terms of the mythology which informs participants' activities. A particular curve, for example, might constitute a breakthrough; or a sheet of figures can count as clear support for some previously postulated theory.

As we have already indicated, however, the cultural specificity of the laboratory does not reside in the mythology available to participants. After all, similar mythologies are available in other laboratories. Specific to this laboratory is the particular configurations of apparatus that we have called inscription devices. The central importance of this material arrangement is that none of the phenomena "about which" participants talk could exist without it. Without a bioassay, for example, a substance could not be said to exist. The bioassay is not merely a means of obtaining some independently given entity; the bioassay constitutes the construction of the substance. Similarly, a substance could not be said to exist without fractionating columns (Photograph 7), since a fraction only exists by virtue of the process of discrimination. Likewise, the spectrum produced by a nuclear magnetic resonance (NMR) spectrometer (Photograph 8) would not exist but for the spectrometer. It is not simply that phenomena *depend on* certain material instrumentation; rather, the phenomena *are thoroughly constituted by* the material setting of the laboratory. The artificial reality, which participants describe in terms of an objective entity, has in fact been constructed by the use of inscription devices. Such a reality, which Bachelard (1953) terms the "phenomeno-technique," takes on the appearance of a phenomenon by virtue of its construction through material techniques.

It follows that if our observer was to imagine the removal of certain items of equipment from the laboratory, this would entail the removal of at least one object of reality from discussion. This was particularly apparent whenever equipment broke down or whenever new equipment was brought into the laboratory.[14] Obviously, however, not all pieces of equipment condition the existence of phenomena and the production of papers in the same way. Taking away the trash can, for example, would be unlikely to harm the main research process; similarly, withdrawal of the automatic pipette would not prevent pipetting by hand, even though this takes longer. By contrast, if the gamma counter breaks down, it is difficult to measure amounts of radioactivity merely by sight! The observation of radioactivity is entirely dependent on the counter (Yalow and Berson, 1971). Clearly, the laboratory would stop operating without the pipes carrying water and oxygen which run between the laboratory and the plant (Photograph 1), but they do not account for the fact that the laboratory produces papers. Like Aristotle's notion of vegetative life, these pipes are a general condition of a superior life but they do not explain it. However, whereas Photograph 1 could have been taken in any factory setting, Photograph 3 is, by contrast, peculiar to a laboratory. This is because apart from the hair dryer, electric motor, and two hydrogen bottles, all the other pieces of apparatus were invented specifically to assist in the construction of laboratory objects. The centrifuge (on the left of Photograph 3), for example, was devised by Svedberg in 1924 and was responsible for creating the notion of protein by allowing undifferentiated substances to be discriminated by spinning (Pedersen, 1974). The molecular weight of proteins could hardly be said to exist except by virtue of the ultracentrifuge. The Rotary Evaporator (on the right of Photograph 3), invented by Craig at the Rockefeller Institute in 1950 (Moore, 1975), enables the removal of solvents in most laboratory purification processes and superceded the previous use of the Claisen flask.

It is clear, then, that some items of equipment are more crucial to the research process than others. Indeed the strength of the laboratory depends not so much on the availability of apparatus, but on the presence of a particular configuration of machines specifically tailored for a particular task. Photograph 3 does not define the particular field in which the work of the laboratory is situated because centrifuges and rotary evaporators can be found in a wide variety of biologically inclined research institutions. However, the presence of bio- and radioimmuno-assays, the Sephadex columns, and the whole gamut of spectrometers, show that participants are concerned with neuro-endocrinology. A whole range of inscription devices, variously used to make points in different subfields, have been assembled in one place. The mass spectrometer, for example, is used in the production of papers on the structure of a substance; cell cultures are used to make points about the synthesis of DNA in the biosynthesis of the same substances.

The cultural specificity of the laboratory is also evident from the fact that some of its inscription devices can only be found in this setting. Most of the substances depend for their existence on bio- and radioimmuno-assays. Each assay comprises several hundred sequences, and sometimes occupies two or three people full time for several days or weeks at a stretch. The instructions for one assay (the TRF immunoassay) occupy six full pages and read like a complicated recipe. Since only relatively small steps, such as pipetting, can be automated, the process relies heavily on the routinised skills of the technicians. As a whole, the assay is an idiosyncratic process in that it depends on the skills of individual technicians and on the use of particular antisera, which themselves have to be obtained from particular goats at particular times of the year. This is why so many substances exist only *locally* (see Chapter 4). The presence in this setting of what scientists refer to as "an exquisite bioassay for growth hormones" or of a "very sensitive assay for CRF" is highly valued by members, and is the source both of their pride as well as the points they make in the literature.

It would be wrong to contrast the material with conceptual components of laboratory activity. The inscription devices, skills, and machines which are now current have often featured in the past literature of *another field*. Thus, each sequence of actions and each routinised assay has at some stage featured as the object of debate in another field and has been the focus of several published papers. The apparatus and craft skills present in one field thus embody the end results of debate or controversy in some other field and make these results available within the walls of the laboratory. It is in this sense that Bachelard (1953) referred to apparatus as "reified theory." The inscription device provides inscriptions which can be used to write papers or to make points in the literature on the basis of a transformation of established arguments into items of apparatus. This transformation, in turn, allows the generation of new inscriptions, new arguments and potentially new items of apparatus (cf., Chapter 6). When, for example, a member of the laboratory uses a computer console (Photograph 11), he mobilises the power of both electronics and statistics. When another member handles the NMR spectrometer (Photograph 8) to check the purity of his compounds, he is utilising spin theory and the outcome of some twenty years of basic physics research. Although Albert knows little more than the general principles of spin theory, this is sufficient to enable him to handle the switchboard of the NMR and to have the power of the theory working to his advantage. When others discuss the spatial structure of a releasing factor, they implicitly make use of decades of research in elementary chemistry. Similarly, a few principles of immunology and a general knowledge of radioactivity are sufficient to benefit from these two sciences when using the radioimmunoassay in the quest for a new substance (Yalow and Berson, 1971). Every move in the laboratory thus relies in some way on other scientific fields. In Table 2.2 we list some of the larger items of equipment used in the laboratory, together with the field of origin and the date at which they were imported into the new problem area. In the next chapter we shall see why much of this equipment originated in fields thought to be "harder" than endocrinology.

Table 2.2

Name of Instrument	Date of First Conception	Date of First Introduction	Field of Origin	Usage in the Program	Remarks
Mass spectrophotometer	1910-1924	1959 for peptides, 1969 for releasing factors	physics (isotopes)	first programme	one PhD to operate it; takes one room
Nuclear magnetic resonance (high resolution) spectrometer	1937-1954	1957 for peptides (pep.), 1964 for releasing factors (R.F.)	physics (spin)	second programme, task one	used to check purity
Amino acid analyzer	1950-1954	within peptide chemistry	protein chemistry; analytic	first and second programmes	routine; machine; automatised
Peptide automatic synthetizer	1966	within pep. for R.F. 1975	biochemistry; synthetic	second programme, task one	routine; machine; automatised, new
Sephadex columns	1956-1959	1960-1962 for R.F.		first, second, and third programmes	essential part of purification and assays
Radioimmunoassay	1956-1960	1959 for pep.	nuclear physics; immunology; endocrinology	all programmes	most versatile and labour intensive instrument
High pressure liquid chromatograph	1958-1967	1973 for pep. 1975 for R.F.	analytical chemistry	first programme and second programme, task one	new, transformed in routine task
Countercurrent distribution chromatograph	1943-1947	1958 for R.F.	"	"	cold piece of machinery

67

Since the material setting represents the reification of knowledge established in the literature of another field, there is necessarily a time lag between the discussion of a theory in one field and the appearance of a corresponding technique in another. This is confirmed by the dates of first conception of various inscription devices. In general, inscription devices were derived from a well-established body of knowledge. Chromatography, for example, is still an active research area of chemistry. But the chromatography embodied in apparatus used in the laboratory dates from Porath's work in the 1950s (Porath, 1967). The mass spectrometer, a crucial analytical tool, is based on physics which is some fifty years old (Beynon, 1960). The same is the case for the laboratory's use of statistics and programming techniques. By borrowing well-established knowledge, and by incorporating it in pieces of furniture or in routine operational sequences, the laboratory can harness the enormous power of tens of other fields for its own purposes.

However, the accumulation of material theories and practices from other fields itself depends on certain manufacturing skills. For example, the mere existence of a discipline such as nuclear physics does not in itself ensure the presence of a beta-counter in the laboratory. Clearly, the use of such equipment presupposes their manufacture. Without Merrifield's invention, for example, there would be no solid phase synthesis and no way of automating peptide synthesis (Merrifield, 1965; 1968). But even without a company like Beckmann, there would still be a prototype at the Rockefeller Institute where it was invented and this could be used by other scientists. Apart from the automatic pipette, a simple time-saving device, both the principle and basic prototype of all the other apparatus used in the laboratory originated in other scientific laboratories. However, industry plays an important role in designing, developing, and making these scientific prototypes available to a larger public, as is clear if we imagine that there were only one or two existing prototypes of each item of new equipment. In this case, scientists would have to travel vast distances and there would be a dramatic fall in the rate of production of papers. The transformation of Merrifield's original prototype into the marketable, self-contained, reliable, and compact item of equipment sold under the name of Automatic Peptide Synthesizer, is a measure of the debt of the laboratory to technological skills (Anonymous, 1976a). If inscription devices are the reification of theories and practices, the actual pieces of equipment are the marketed forms of these reifications.

The material layout of the laboratory has been constructed from items of apparatus, many of which have long and sometimes controversial histories. Each item of apparatus has combined with certain skills to form specific devices, the styluses and needles of which scratch the surface of sheets of graph paper. The string of events to which each curve owes it very existence is too long for any observer, technician, or scientist to remember. And yet each step is crucial, for its omission or mishandling can nullify the entire process. Instead of a "nice curve," it is all too easy to obtain a chaotic scattering of random points of curves which cannot be replicated. To counter these catastrophic possibilities, efforts are made to routinise component actions either through technicians' training or by automation. Once a string of operations has been routinised, one can look at the figures obtained and quietly forget that immunology, atomic physics, statistics, and electronics actually made this figure possible. Once the data sheet has been taken to the office for discussion, one can forget the several weeks of work by technicians and the hundreds of dollars which have gone into its production. After the paper which incorporates these figures has been written, and the main result of the paper has been embodied in some new inscription device, it is easy to forget that the construction of the paper depended on material factors. The bench space will be forgotten, and the existence of laboratories will fade from consideration. Instead, "ideas," "theories," and "reasons" will take their place. Inscription devices thus appear to be valued on the basis of the extent to which they facilitate a swift transition from craft work to ideas. The material setting both makes possible the phenomena and is required to be easily forgotten. Without the material environment of the laboratory none of the objects could be said to exist, and yet the material environment very rarely receives mention. It is this paradox, which is an essential feature of science, that we shall now consider in more detail.

Documents and Facts

Thus far, our observer has begun to make sense of the laboratory in terms of a tribe of readers and writers who spend two-thirds of their time working with large inscription devices. They appear to have developed considerable skills in setting up devices which can pin down elusive figures, traces, or inscriptions in their craftwork, and in the art of persuasion. The latter skill enables them to convince others that what they do is important, that what they say is true, and that their proposals are worth funding. They are so skillful, indeed, that they manage to convince others not that they are being convinced but that they are simply following a consistent line of interpretation of available evidence. Others are persuaded that they are not persuaded, that no mediations intercede between what is said and the truth. They are so persuasive, in fact, that within the confines of their laboratory it is possible to forget the material dimensions of the laboratory, the bench work, and the influence of the past, and to focus only on the "facts" that are being pointed out. Not surprisingly, our anthropological observer experienced some dis-ease in handling such a tribe. Whereas other tribes believe in gods or complicated mythologies, the members of this tribe insist that their activity is in no way to be associated with beliefs, a culture, or a mythology. Instead, they claim to be concerned only with "hard facts." The observer is puzzled precisely because his informants insist that everything is straight-forward. Moreover, they argue that if he were a scientist himself, he would understand this. Our anthropologist is sorely tempted by this argument. He has begun to learn about the laboratory, he has read lots of papers and can recognise different substances. Furthermore, he begins to understand fragments of conversation between members. His informants begin to sway him. He begins to admit that there is nothing strange about this setting and nothing which requires explanation in terms other than those of informants' own accounts. However, in the back of his mind there remains a nagging question. How can we account for the fact that in any one year, approximately one and a half million dollars is spent to enable twenty-five people to produce forty papers?

Apart from the papers themselves, of course, another kind of product provides the means for generating documents in other laboratories. As we said above, two of the main objectives of this laboratory are the purification of natural substances and the manufacture of analogs of known substances. Frequently, purified fractions and samples of synthetic substances are sent to investigators in other laboratories. Each analog is produced at an average cost of $1,500, or $10 per milligram, which is much lower than the market value of these peptides. Indeed, the market value of all peptides produced by the laboratory would amount to $1.5 million, the same as the total budget of the laboratory. In other words, the laboratory could pay for its research by selling its analogs. However, the quantities, the number, and the nature of the peptides actually produced by the laboratory are such that there is no market for 99 percent of them. Moreover, nearly all the peptides (90 percent) are manufactured for internal consumption and are not available as output. The actual output (for example, 3.2 grams in 1976) is potentially worth $130,000 at market value, and although it cost only $30,000 to produce, samples are sent free of charge to outside researchers who have been able to convince one of the members of the laboratory that his or her research is of interest. Although members of the laboratory do not require their names to appear on papers which report work resulting from the use of these samples, the ability to provide rare and costly analogs is a powerful resource. If, for example, only a few micrograms were made available, this would effectively prevent the recipient from carrying out sufficient investigations to make a discovery (see Chapter 4).[15] Purified substances and rare antisera are also considered valuable assets. When, for example, a participant talks about leaving the group, he often expresses concern about the fate of the antiseras, fractions, and samples for which he has been responsible. It is these, together with the papers he has produced, that represent the riches needed by a participant to enable him to settle elsewhere and write further papers. He is likely to find similar inscription devices elsewhere, but not the idiosyncratic antisera that permit a specific radioimmunoassay to be run. Besides samples, the laboratory also produces skills in the members of a workforce who from time to time leave the laboratory to work elsewhere. Here again, the skill is only a means to the end of publishing a paper.

The production of papers is acknowledged by participants as the main objective of their activity. The realisation of this objective necessitates a chain of writing operations from a result first scribbled on a sheet of paper and enthusiastically communicated to colleagues, to the final registering of published literature in the laboratory archives. The many intermediary stages (such as talks with slides, circulation of preprints, and so on) all concern literary production of one kind or another. It is thus necessary carefully to study the various processes of literary production which lead to the output of papers. We shall do this in two ways. Firstly, we shall consider papers as objects in much the same way as manufactured goods. Secondly, we shall attempt to make sense of the content of papers. By looking at literary production in this way we hope to broach the central questions posed by our observer: how can a paper be both so expensive to produce and yet so highly valued? What exactly can justify participants' faith in the importance of the papers' contents?

THE PUBLICATION LIST

The range and scope of papers produced by the laboratory is indicated by a list kept and updated by participants. We used those items listed between 1970 and 1976. Although referred to by participants as the "publication list," a number of articles were included which had not in fact been published.[16]

Let us classify output according to the channel chosen by investigators. Fifty percent consisted of "regular" papers. Such items comprised several pages and were published in professional journals. Twenty percent of the output comprised abstracts submitted to professional congresses. A further 16 percent comprised solicited contributions to meetings, only half of which found their way into print as conference proceedings. Participants also contributed chapters to edited collections of papers, which made up 14 percent of the total output.

Another way of classifying papers is by the literary "genre" of articles. Differences in genre were defined both in terms of formal characteristics (such as the size, style, and format of each article) and by the nature of the audience. For example, 5 percent of all papers were addressed to lay audiences, such as lay readers of *Scientific American, Triangle,* and *Science Year* or to physicians for whom a simplified account of recent progress in biology is available in articles, such as those in *Clinician, Contraception,* or *Hospital Practice.* Although a relatively minor output in terms of quantity, this genre fulfills an important public relations function in that such articles can be useful in the long-term acquisition of public funds. A second genre, which made up 27 percent of total output, addressed scientists working outside the releasing factors field. Sample titles included: "Hypothalamus Releasing Hormones," "Physiology and Chemistry of the Hypothalamus," and "Hypothalamic Hormones: Isolation, Characterisation and Structure Function." The details of specific substances and assays or of the relations between them were rarely discussed in these kinds of articles, which could be found most frequently in advanced textbooks, reference books, nonspecialised journals, book reviews, and invited lectures. The information in these articles was often utilised by students or by colleagues in outside fields. Such papers are both incomprehensible to laymen and unremarkable to colleagues within the field of releasing factors. They simply summarize the state of the art for scientists outside the field. A third genre, which made up 13 percent of the total output, included titles such as: "Luteinizing Releasing Factor and Somatostatin Analogs: Structure Function Relationships," "Biological Activities of SS," and "Chemistry and Physiology of Ovine and Synthetic TRF and LRF." These articles were specialised to the extent that they made little sense outside the specialty. They were characterised by an unusually high number of coauthors (5.7 compared with an average of 3.8 for all papers) and were usually presented at professional meetings within the field such as the Endocrine Society Meetings and Peptide Chemistry Symposia. Articles in this third genre enabled colleagues to catch up on the latest available information. Lastly, a genre which made up 55 percent of the total output comprised highly specialised articles as indicated by the following example titles: "(Gly) 2LRF and Des His LRF. The synthesis purification and characterisation of two LRF analogs antagonists to LRF" and "Somatostatin inhibits the release of acetylcholine induced electrically in the myenteric plexus." Such articles, which aimed to convey minute pieces of information to a select band of insiders, were published mainly in journals such as *Endocrinology* (18 percent), *BBRC* (10 percent), and *Journal of Medical Chemistry* (10 percent). Whereas papers falling within the first and second genres were thought to be important in a teaching context, only those articles in the latter two genres (the insider reviews and specialised articles) were regarded by members of the laboratory as containing new information.

By dividing the annual budget of the laboratory by the number of articles published (and at the same time discounting those articles in the laymen's genre), our observer calculated that the cost of producing a paper was $60,000 in 1975 and $30,000 in 1976. Clearly, papers were an expensive commodity! This expenditure appears needlessly extravagant if papers have no impact, and extravagantly cheap if papers have fundamental implications for either basic or applied research. It may therefore be appropriate to interpret this expenditure in relation to the reception of papers.

One preliminary method of examining the cost of production in relation to the received value of papers is through an examination of citation histories. Our observer used the SCI to trace the citations of the 213 items[17] published by participants between 1970 and 1976. Items that were not cited (articles by laymen, unpublished lectures, and abstracts that were difficult to obtain) were then weeded out and the remainder divided into those highly likely to be cited and those (usually chapters of books or abstracts) that were not. Since the peak of citation activity rarely occurred later than the fourth year following publication, the observer calculated an index of each item's impact based on citations in the year of publication and in the subsequent two years.

The overall impact ratio (number of citations per item) was 12.4 c.p.i. for the five years for which it could be calculated (1970-1974). However, this figures conceals three main sources of variation. Firstly, impact ratio varied according to genre. For example, when only "regular" papers were considered, the impact ratio rose to 20 c.p.i. Furthermore, only 17 of the items identified as "regular" papers and published in what participants referred to as "good" journals had no impact whatsoever before the end of 1976. Secondly, impact ratio varied over time. It was 23.2 c.p.i. for the 10 items published in 1970, but only 8 c.p.i. for the 39 items published in 1974. This particular variation is explained by the fact that 1970 was the year of a major discovery (see Chapter 3). Thirdly, as is evident from the right-hand column of Table 2.1, impact ratio also varied by programme. Of the three programmes we characterised earlier, those items concerning the isolation and characterisation of substances had the highest impact ratio (24 c.p.i.). Only one other category of activity, production of analogs carried out in collaboration with clinicians (task three of the second programme), had comparable impact (21 c.p.i.). Items resulting from other activities had much less impact. The third programme, for example, made up 22 percent of overall output (in terms of items produced) but had an impact ratio of only 10.6 c.p.i. Task two of the second programme made up a similar proportion of overall output (24 percent) but had even less impact (7.6 c.p.i.).

If impact ratio is taken as a crude indicator of return on the initial costs of producing items of literature, it is clear that a higher level of return is not necessarily guaranteed by increased output. One dominant factor would appear to be the extent to which items can appear as "regular" papers. However, this is confused both by variations over time and by the particular activity associated with each item. We are left, therefore, with the somewhat tautological speculation that items which yield a high return are those with a high chance of addressing issues of concern outside the laboratory.

STATEMENT TYPES

Although citations revealed that items had varying impact, our observer felt that he had discovered little about why this was the case. One reaction to this kind of problem is to engage in more sophisticated and complex mathematical analysis of citation histories, in the hope that some clearly identifiable pattern of citations will emerge.[18] But our observer was unconvinced that this would alleviate his basic difficulty of understanding why items were cited in the first place. Instead, he reasoned that there must be something in the *content* of papers which would explain how they were evaluated. Accordingly, our observer began to peruse some of the articles in order to ferret out possible reasons for their relative value. Alas, it was all Chinese to him! Many of the terms were recognisable as the names of substances, or of apparatus and chemicals which he had already come across. He also felt that the grammar and the basic structure of sentences was not dissimilar to those he used himself. But he felt entirely unable to grasp the "meaning" of these papers, let alone understand how such meaning sustained an entire culture. He was reminded momentarily of an earlier study of religious rituals when, having penetrated to the core of ceremonial behaviour, he had found only twaddling and waffling. In a similar way, he had now discovered that the end products of a complex series of operations contained complete gibberish. In desperation he turned to participants. But his requests for clarification of the meaning of papers were met with retorts that the papers had no interest or significance *in themselves*: they were only a *means* of communicating "important findings." When further asked about the nature of these findings, participants merely repeated a slightly modified version of the content of the papers. They argued that the observer was baffled because his obsessive interest in literature had blinded him to the real importance of the papers: only by abandoning his interest in the papers themselves could the observer grasp the "true meaning" of the "facts" which the paper contained.

Our observer might have become extremely depressed by participants' scorn, were it not for the fact that participants immediately resumed their discussion of drafts, the correction and recorrection of galley proofs, and the interpretation of various traces and figures which had just been produced by inscription devices. At the very least, reasoned our observer, there must be a strong relationship between processes of literary inscription and the "true meaning" of papers.

The above disagreement between observer and participant hinged on a paradox which had already been hinted at several times during this chapter. The production of a paper depends critically on various processes of writing and reading which can be summarised as literary inscription. The function of literary inscription is the successful persuasion of readers, but the readers are only fully convinced when all sources of persuasion seem to have disappeared. In other words, the various operations of writing and reading which sustain an argument are seen by participants to be largely irrelevant to "facts," which emerge solely by virtue of these same operations. There is, then, an essential congruence between a "fact" and the successful operation of various processes of literary inscription. A text or statement can thus be read as "containing" or "being about a fact" when readers are sufficiently convinced that there is no debate about it and the processes of literary inscription are forgotten. Conversely, one way of under-cutting the "facticity" of a statement is by drawing attention to the (mere) processes of literary inscription which make the fact possible. With this in mind, our observer decided to look carefully at the different kind of statements to be found in the papers. In particular, he was concerned to delineate the extent to which some statements appeared more fact-like than others.

At one extreme, readers are so persuaded of the existence of facts that no explicit reference is made to them. In other words, various items of knowledge are simply taken for granted and utilised in the course of an argument whose main burden is the explicit demonstration of some other fact. Consequently, it was difficult when reading articles consciously to note the occurrence of taken-for-granted facts. Instead, they merged imperceptibly into a background of routine enquiry, skills, and tacit knowledge. It was obvious to our observer, however, that everything taken as self-evident in the laboratory was likely to have been the subject of some dispute in earlier papers. In the intervening period a gradual shift had occurred whereby an argument had been transformed from an issue of hotly contested discussion into a well-known, unremarkable and noncontentious fact. The observer therefore posited a five-fold classificatory scheme corresponding to different types of statements. Statements corresponding to a taken-for-granted fact were denoted *type* 5 statements. Precisely because they were taken for granted, our observer found that such statements rarely featured in discussions between laboratory members, except when newcomers to the laboratory required some introduction to them. The greater the ignorance of a newcomer, the deeper the informant was required to delve into layers of implicit knowledge, and the farther into the past. Beyond a certain point, persistent questioning by the newcomer about "things that everybody knew" was regarded as socially inept. In the course of one discussion, for example, X repeatedly argued that "in the grid test rats do not react as if they were on neuroleptics." For X, the force of the argument was clear. But for Y, a scientist working in a different field, there were preliminary questions to ask: "What do you mean by a grid test?" Somewhat taken aback, X stopped, looked at Y, and adopted the tone of a teacher reading from a textbook: "The classic catalepsy test is a vertical screen test. You have a wire mesh. You put the animal on the wire mesh and an animal which has been injected with neuroleptic will remain in this position. An animal which is untreated, will just climb down" (IX, 83). For X, his earlier reference to the assay was a type 5 statement which required no further explication. After this interruption, X adopted his previous excited tone and returned to the original argument.

Scientific textbooks were found to contain a large number of sentences with the stylistic form: "A has a certain relationship with B." For example, "Ribosomal proteins begin to bind to pre-RNA soon after its transcription starts" (Watson, 1976: 200). Expressions of this sort could be said to be *type 4* statements. Although the relationship presented in this statement appears uncontroversial, it is, by contrast with *type 5* statements, made explicit. This type of statement is often taken as the prototype of scientific assertion. However, our observer found this type of statement to be relatively rare in the work of scientists in the laboratory. More commonly, *type 4* statements formed part of the accepted knowledge disseminated through teaching texts.

Another kind of statement consisted of expressions with the form, "A has a certain relationship with B," which were embedded in other expressions: "It is still largely unknown which factors cause the hypothalamus to withhold stimuli to the gonads" (Scharrer and Scharrer, 1963). "Oxytocin is generally assumed to be produced by the neurosecretory cells of the paraventricular nuclei" (Olivecrona, 1957; Nibbelink, 1961). These were referred to as *type 3* statements. They contained statements about other statements which our observer referred to as *modalities*.[19] By deleting modalities from *type 3* statements it is possible to obtain *type 4* statements. The difference between statements in textbooks and the above, many of which appeared in review articles (Greimas, 1976), can thus be characterised by the presence or absence of modalities. A statement clearly takes on a different form when modalities drop. Thus, to state, "The structure of GH.RH was *reported to be* X" is not the same as saying, "The structure of GH.RH *is* X." Our observer found many different types of modality. One form of statement, for example, included a reference and a date in addition to the basic assertion. In other statements, modalities comprised expressions relating to the merit of the author or to the priority of work which had initially postulated the relationship in question: "[T]his method has *first* been described by Pietta and Marshall. Various investigators clearly established [ref.] . . . ." "More convincing evidence was provided by [ref.] . . . ." "[T]he first unequivocal demonstration was provided by [ref.] . . . ." (all quotations from Scharrer and Scharrer, 1963).

As mentioned above, many *type 3* statements were found in review discussions. Much more common among the papers and drafts circulated in the laboratory were statements which appeared rather more contentious than those in reviews.

> Recently Odell [ref.] has reported that hypothalamic tissues, when incubated . . . woud increase the amount of TSH. It is difficult to ascertain whether or not . . . .

> At this time we do not know whether the long acting effect of these compounds extract to their potential inhibitory activity (Scharrer and Scharrer, 1963).

Statements of this form appeared to our observer to more nearly constitute *claims* rather than established facts. This was because the modalities which encompassed expressions of basic relationships seemed to draw attention to the circumstances affecting the basic relationship. Statements containing these kinds of modalities were designated *type 2* statements. For example:

> There is a large body of evidence to support the concept of a control of the pituitary by the brain.

> The role of nitrogen 1 and nitrogen 3 of the imidazole ring of histidine in TRF and LRF seems to be different.

> It is unlikely that racemization occurs during esterification with any of the above procedures, but little experimental evidence is available to support this point (Scharrer and Scharrer, 1963).

More precisely, *type 2* statements could be identified as containing modalities which draw attention to the generality of available evidence (or the lack of it). Basic relationships are thus embedded within appeals to "what is generally known" or to "what might reasonably be thought to be the case." The modalities in *type 2* statements sometimes take the form of tentative suggestions, usually oriented to further investigations which may elucidate the value of the relationship at issue:

> It should not be forgotten that hypothalamic tissues contain non-negligible quantities of TSH . . . which may further complicate the interpretation of the data. . . It would be interesting to ascertain whether or not their material is similar. . . . It is somewhat puzzling that . . . (Scharrer and Scharrer, 1963).

Type 1 statements comprise conjectures or speculations (about a relationship) which appear most commonly at the end of papers, or in private discussions:

> Peter [ref.] has suggested that in goldfish the hypothalamus has an inhibitory effect on the secretion of TSH.
> There is also this guy in Colorado. They claim that they have got a precursor for H . . . . I just got the preprint of their paper (III, 70).
> It may also signify that not everything seen, said and reasoned about opiates may necessarily be applicable for the endorphins.

By this stage, then, our observer had identified five different types of statement. At first glance it seemed that these types could be arranged in a broad continuum such that *type 5* statements represented the most fact-like entities and *type 1* the most speculative assertions. It would follow that changes in statement type would correspond to changes in fact-like status. For example, the deletion of modalities in a *type 3* statement would leave a *type 4* statement, whose facticity would be correspondingly enhanced. At a general level, the notion that changes in statement type may correspond to changes in facticity seems plausible enough. At the level of empirical verification, however, this general scheme encounters certain difficulties.

In any given instance, there seems to be no simple relationship between the form of a statement and the level of facticity which it expresses. This can be demonstrated, for example, by considering a statement which contains an assertion about the relationship between two variables together with a reference. As it stands, our observer would classify this statement as a *type 3* where the modality is constituted by the included reference. Undoubtedly, the deletion of the modality would leave a *type 4* statement. It is questionable, however, whether this would enhance or detract from the fact-like status of the statement. On the one hand, we could argue that the inclusion of a reference draws attention to circumstances surrounding the establishment of the relationship in question and that this, by implication, renders the relationship less indisputable and hence less likely to be taken for granted. By noting that human agency was involved in its production, the inclusion of a reference diminishes the likelihood that the statement will be accepted as an "objective fact of nature." On the other hand, it could be argued that the inclusion of a reference lends weight to a statement which otherwise appears to be an unsupported assertion. Thus, it is only by virtue of the reference that the statement achieves any degree of facticity.

The determination of the correct or more appropriate interpretation of the function of a modality will depend critically on our knowledge of the context in each particular case. If, for example, we have good grounds for supposing that the inclusion of a modality in a paper was a presentational device designed to enhance the acceptance of a statement, then the onus is upon us to provide details of the context in which this device was so used. There are, of course, those who argue that this kind of determinate relationship between context and a particular interpretation of a statement simply does not exist. For our purposes, however, it is sufficient to note that changes in the type of statement provide the *possibility* of changes in the fact-like status of statements. Even though, in any individual instance, we may not be able unambiguously to specify the direction of change in facticity, we retain the possibility that such changes *can* correspond to changes in statement types.

Because he was aware of the problems both of specifying the fact-like status of any given statement and of specifying the direction of change of facticity in any example, our observer felt he could not stake a great deal on the determinacy of correspondence between statement type and fact-like status. Nevertheless, he realised that the notion of literary inscription had provided a useful tool. Although he understood little of the content of the papers he was reading, he had developed a simple grammatical technique for distinguishing between types of statements. This, he felt, enabled him to approach the very substance of scientists' statements without having entirely to rely on participants for elucidation or assistance. Furthermore, to the extent that changes in the grammatical form of scientists' statements provided the possibility of changes in their content (or fact-like status), he could portray laboratory activity as a constant struggle for the generation and acceptance of particular types of statement.

THE TRANSFORMATION OF STATEMENT TYPES

Despite the simplicity of the classificatory scheme presented above (and summarized in Figure 2.3), it at least provided our anthropologist with a tentative means of ordering his observations of the laboratory which was consistent with his earlier notion of literary inscription. Activity in the laboratory had the effect of transforming statements from one type to another. The aim of the game was to create as many statements as possible of *type 4* in the face of a variety of pressures to submerge assertions in modalities such that they became artefacts. In short, the objective was to persuade colleagues that they should drop all modalities used in relation to a particular assertion and that they should accept and borrow this assertion as an established matter of fact, preferably by citing the paper in which it appeared. But how precisely is this achieved? What exactly are the operations which successfully transform statements?

Consider the following example, in which John interrupts K's description of an assay in which the effect of LH had apparently been blocked.

John: Since melatonin inhibits LH we cannot be sure that you are not simply measuring melatonin.

K: I don't believe these data on the release of LH by melatonin . . . not in my system (VI, 18).

Instead of simply accepting K's previous statement, John adds a modality ("we can not be sure") to the unstated assumption that the investigators were "not simply measuring melatonin." John thus casts doubt on an original unstated, and hence *type 5* statement by using a qualification about the consensual certainty which investigators ("we") are entitled to assume. As a result, the original *type 5* statement is transformed into a highly conjectural *type 2* statement. The transformation is made particularly effective in this case by the preceding justification for investigator's lack of sureness. "Since melatonin inhibits LH" constitutes the use of a *type 4* statement to justify the addition of a modality to the originally unstated assumption. K's response attempts to recast John's justificatory *type 4* statement

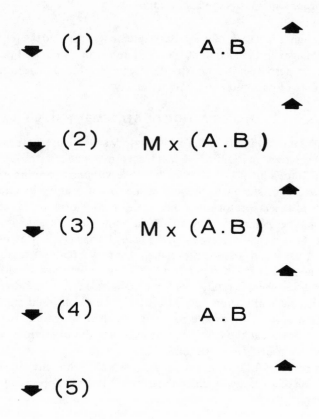

Figure 2.3
This diagram represents the different stages a statement—A.B—undergoes before becoming a fact. A fact is nothing but a statement with no modality—M—and no trace of authorship. The last stage—5—characterises the implicit dimension of something so obvious that it does not even have to be said. To move a statement from one stage to another, operations have to be performed. As indicated by the arrows, a given statement may move toward a fact-like status—from 1 to 5—or toward an artefact-like status—from 5 to 1—(see Ch. 4).

by adding a modality. By "not believing" circumstances surrounding the establishment of "melatonin inhibits LH," K tries to undercut John's attempt to undercut the unstated assumption that "you are not simply measuring melatonin."

A second example is an excerpt from a paper written by John: "Our original observations (ref.) of the effects of somatostatin on the secretion of TSH have now been confirmed in other laboratories (ref.)." John had written an earlier paper, to which he first refers, and the statements contained therein had been subsequently confirmed. Whereas the statement, "the effects of somatostatin on the secretion of TSH," had originally appeared as a claim of *type 2*, it now appears as an assertion embedded within references and enhanced by the modality "have now been confirmed." In this way, John was able to borrow a statement made by others in order to transform his own initial statement into *type 3*.

The above examples demonstrate the use of two related operations. The first effects a change in the existing modality which can either enhance or detract from the facticity of a given statement. The second borrows an existing statement type in such a way that its facticity can be either enhanced or diminished (Latour, 1976).

The observer was now able to think of what had previously appeared a confused mixture of papers in terms of a network of texts containing a multitude of statements. The network itself comprised a large body of operations on and between these statements. It would thus be possible to document the history of a particular assertion as it became transformed from one statement type into another and as its factual status was continually diminished or enhanced as the result of various operations. We have already specified, in a preliminary way, the nature of operations by which statement types becomes transformed. Let us now examine in more detail one criterion for the success of an operation.

Our observer recalled that the inscriptions produced by certain configurations of apparatus were "taken seriously" if they could be read as being the same as other inscriptions produced under the same conditions. In simple terms, participants were more convinced that an inscription unambiguously related to a substance "out there," if a similar inscription could also be found. In the same way, an important factor in the acceptance of a statement was the recognition by others of another statement which was similar. The combination of two or more apparently similar statements concretised the existence of some external object or objective condition of which the statements were taken to be indicators. Sources of "subjectivity" thus disappeared in the face of more than one statement, and the initial statement could be taken at face value and without qualification (cf., Silverman, 1975). It is in this manner that our scientists, when noticing a peak on the spectrum of a chromatograph, sometimes rejected it as noise. If, however, the same peak was seen to occur more than once (under what were regarded an independent circumstances), it was often said that there was a substance there of which the peaks were a trace. An "object" was thus achieved through the superimposition of several statements or documents in such a way that all the statements were seen to relate to something outside of, or beyond, the reader's or author's subjectivity.[20] Similarly, the introduction, or rather the *re*introduction, of an author's subjectivity as essentially linked to the production of a statement could be used to diminish the factual status of the statement. In the laboratory, "objects" were accomplished by the superimposition of several documents obtained from inscription devices within the laboratory or from papers by investigators outside the laboratory (cf., Chapter 4). No statement could be made except on the basis of available documents; statements were thus loaded with documents and modalities which constituted an evaluation of the statement. Consequently, grammatical modalities ("maybe," "definitely established," "unlikely," "not confirmed") often acted like price tags of statements, or, to use a mechanical analogy, like an expression of the *weight* of a statement. By adding or withdrawing layers of documents, scientists could increase or decrease qualifications and hence the weight of the statement was modified accordingly. For example, one referee's report included the following: "The conclusion that the effect of Pheno . . . [to] release PRL *in vivo* is mediated through the hypothalamus *is premature.*" Three references were then given, which further pulled the rug from under the author's conclusion. Thus, although the author had presented his statement as a *type 2* or *3*, the referee recast it in terms of *type 1*. Consider also the following: "The authors used a Polytron which is a much more vigorous means of tissue disruption. To my knowledge, *there are no reports in the literature* of successful subcellular fractionation of brain tissue disruption." In this case the referee cast doubt on the use of a machine which produced the documents on which the argument is based. This was done by reference to a notable absence of any statements which might justify and hence enhance the authors' original claim. As a result, the authors' (unsupported) claim must be read in conjunction with diminishing modalities such as "there is no support for this" and is consequently to be regarded as worthless.

With the notion of operations between (and on) statements in the literature, our observer began to feel more confident in his ability to understand the layout of individual papers. As a brief indication of the scope of the analysis which this permitted, let us look closely at one of the papers produced by the laboratory (Latour, 1976; Latour and Fabri, 1977).

The introductory paragraph refers to four articles, previously published by members of the laboratory, in which they posited the structure of a particular substance B. This referencing can be read as the invocation of documents which bear upon the present problem. More specifically, the use of these past papers can be read as providing support for the present enterprise. (The grounds for this particular reading are simply that the four papers themselves received 400 citations, all of which appear confirmatory.) At the same time, however, the papers are themselves taken as statement *type 3*, for which further support is to be provided by the present argument: "this short note reports data obtained in rats which *confirm and expand our early results.*" The three following paragraphs summarize the way in which inscription devices were set up so as to obtain data. The information appears here in the form of *type 5* statements. In other words, knowledge is invoked which is so common to an audience of potential readers that no citations are necessary: "All synthetic preparations of substance B had full biological activity as ascertained in 4 or 6 point assays *in vitro* with factorial analysis."

In each of the next statements from the "results" section of the paper, reference is made to a figure.

"The results shown in Fig. 2 demonstrate that substance B significantly lowers blood levels of GH for 20 to 40 mn but not for 40 to 50 mn." Each figure thus acts as a tidied representation of documents (obtained from a radioimmunoassay) which is used in the text to support a particular point. It is not simply that "the results demonstrate that . . . ." Rather, these results have an external reference and an independent existence which can be supported by the presence of "Fig. 2." The inclusion of "shown in Fig. 2" can thus provide an enhanced reading of an otherwise unsupported claim about the results. Subsequent discussion comprises three paragraphs, which refer back to the former "results" section ("These experiments show that . . . .").

The "results" section is itself based on figures which are, in turn, dependent on the inscription devices described earlier. The result of this accumulation of back references is an impression of objectivity: the "fact" that "synthetic substance B inhibits GH in rats" can be taken by the reader as independent of the author's subjectivity and thus worthy of belief.

At the same time the establishment of one statement opens up discussion of others: "The mechanisms of action of the barbiturate in . . . are not well understood." The modality "are not well understood" is not intended to diminish some prior claim about "the mechanisms of action of the barbiturate." Instead, its inclusion in this context amounts to a tentative suggestion for areas of future work. The statement is thus of *type 1* or *2*. As a result, subsequent discussion focuses on this statement as a new proposition: "[*W*]*e might as well envisage* them [the mechanisms] as involving inhibition of secretion of endogenous substance B, a *hypothesis which is not incompatible* with the data." Finally, the new statement is linked to a deontic operation:[21] "This hypothesis will best be approached by some type of radio-immunoassay *still to be developed*."

It should not be forgotten, however, that this paper is itself part of a long series of operations within the field. The SCI shows that between 1974 and 1977 this paper received 62 explicit citations from 53 papers. Of these, 31 appear simply to have borrowed the conclusion (that synthetic substance B inhibits GH as well as natural substance B in the rat) as a fact and used it in their introduction; eight papers focused solely on the final deontic operations in the paper in pursuing the suggestion for further work; two papers by the same author cited the above paper as confirmatory evidence of his own earlier work; and four papers used fresh data further to confirm the original statement. Only one paper raised doubts about the use of the assay in obtaining one of the figures mentioned in the fifth statement ("there are discrepancies between their results and ours"). This one paper examined above thus provided the focus of a variety of operations performed by later articles. Its weight depended both on its use of earlier literature, inscription devices, documents, and statements as well as on subsequent reaction to it.

Conclusion

A laboratory is constantly performing operations on statements; adding modalities, citing, enhancing, diminishing, borrowing, and proposing new combinations. Each of these operations can result in a statement which is either different or merely qualified. Each statement, in turn, provides the focus for similar operations in other laboratories. Thus, members of our laboratory regularly noticed how their own assertions were rejected, borrowed, quoted, ignored, confirmed, or dissolved by others. Some laboratories were seen to be engaged in the frequent manipulation of statements while elsewhere there was thought to be little activity. Some groups produce almost at a loss: they talk and publish, but no one operates on their statements. In such a case, a statement can remain cast as a *type 1*, a claim lingering in an operational limbo. By contrast, other assertions can be seen to change their status rapidly, following a kind of alternate dance, as they are proven, disproven, and proven again. Despite the large number of operations performed on them, they rarely change their form radically. These statements represent a mere fraction of the hundreds of artefacts and half-born statements which stagnate like a vast cloud of smog. Commonly, attention shifts from these to other statements. In some places, however, we can discern a clearer picture. One or other operation irrevocably annihilates a statement never to be taken up again. Or, by contrast, in situations where a statement is quickly borrowed, used and reused, there quickly comes a stage where it is no longer contested. Amidst the general Brownian agitation, a fact has then been constituted. This is a comparatively rare event, but when it occurs, a statement becomes incorporated in the stock of taken-for-granted features which have silently disappeared from the conscious concerns of daily scientific activity. The fact becomes incorporated in graduate text books or perhaps forms the material basis for an item of equipment. Such facts are often thought of in terms of the conditioned reflexes of "good" scientists or as part and parcel of the "logic" of reasoning.

By pursuing the notion of literary inscription, our observer has been able to pick his way through the labyrinth. He can now explain the objectives and products of the laboratory in his own terms, and he can begin to understand how work is organised and why literary production is so highly valued. He can see that both main sections (A and B) of the laboratory are part of the same process of literary inscription. The so-called material elements of the laboratory are based upon the reified outcomes of past controversies which are available in the published literature. As a result, it is these same material elements which allow papers to be written and points to be made. Furthermore, the anthropologist feels vindicated in having retained his anthropological perspective in the face of the beguiling charms of his informants: they claimed merely to be scientists discovering facts; he doggedly argued that they were writers and readers in the business of being convinced and convincing others. Initially this had seemed a moot or even absurd standpoint, but now it appeared far more reasonable. The problem for participants was to persuade readers of papers (and constituent diagrams and figures) that its statements should be accepted as fact. To this end rats had been bled and beheaded, frogs had been flayed, chemicals consumed, time spent, careers had been made or broken, and inscription devices had been manufactured and accumulated within the laboratory. This, indeed, was the very raison d'être of the laboratory. By remaining steadfastly obstinate, our anthropological observer resisted the temptation to be convinced by the facts. Instead, he was able to portray laboratory activity as the organisation of persuasion through literary inscription. Has the anthropologist himself been convincing? Has he used sufficient photographs, diagrams, and figures to persuade his readers not to qualify his statements with modalities, and to adopt his assertions that a laboratory is a system of literary inscription? Unfortunately, for reasons which will later become clear (see Chapter 6), the answer has to be no. He cannot claim to have set forth an account which is immune from all possibility of future qualification. Instead, the best our observer has done is to create a small breathing space. The possibility of future reevaluation of his statements remains. As we shall see in the next chapter, for example, the observer can be forced back into the labyrinth as soon as questions are posed about the historical evolution of any one specific fact.

NOTES

1. We stress that "the observer" is a fictional character so as to draw attention to the process whereby we are engaged in constructing an account (see Chapter 1). The essential similarity of our procedures for constructing accounts and those used by laboratory scientists in generating and sustaining facts will become clear in the course of our discussion. The point is taken up explicitly in Chapter 6.

2. The notion of inscription as taken from Derrida (1977) designates an operation more basic than writing (Dagognet, 1973). It is used here to summarize all traces, spots, points, histograms, recorded numbers, spectra, peaks, and so on. See below.

3. A file of photographs of the laboratory is presented after Chapter 2.

4. See note 2.

5. This notion of inscription device is sociological by nature. It allows one to describe a whole set of occupations in the laboratory, without being disturbed by the wide variety of their material shapes. For example, a "bioassay for TRF" counts as *one* inscription device even though it takes five individuals three weeks to operate and occupies several rooms in the laboratory. Its salient feature is the final production of a figure. A large item of apparatus, such as the Nuclear Magnetic Resonance Spectrometer, is rarely used as an inscription device. It is used instead to monitor a process of peptide production. However, the same apparatus, a scale for instance, can be considered an inscription device when it is used to get information about a new compound; a machine when it is used to weigh some powder; and a checking device when used to verify that another operation has gone according to plan.

6. Our observer was well aware of the popularisation of the term due to Kuhn (1970) and of the subsequent debates over its ambiguity and significance for models of scientific development (see, for example, Lakatos and Musgrave, 1970).

7. We use the term "peptide" throughout the following argument. One classical textbook definition of the peptidic bond is as follows: "A covalent bond between two amino acids in which the alpha amino group of one amino acid is bonded to the alpha-carboxyl group of the other with the elimination of H_2O" (Watson, 1976). In practice, "peptide" is a synonym for a small protein. However, it is important to realise that such terms need *not* be defined as if they have a universal meaning beyond that of the specific culture in which they are used. As if they were the terms used by the tribe under study, we shall enclose such terms in quotes in our discussion and attempt to account for them in nontechnical terms.

8. There are only some twenty amino acids in the body; proteins and peptides are made up exclusively of these amino acids; each amino acid has a name, for example, tyrosine, tryptophene, and proline. In the text we often use a simple abbreviation of these names (which uses the three first letters of the amino acid name).

9. These very crude figures are intended merely to give a general idea of the scale. They are based on the volume of space devoted to different topics in the *Index Medicus*.

10. Once again, these divisions are extremely artificial in that they are much too large and rigid to correspond directly to members' appraisal of their activities. On the other hand, these programmes have become very stable and routinised by comparison with those of other laboratories. Our intention here is merely to provide the reader with the backcloth necessary for understanding subsequent chapters.

11. The observer would be told, for example, that "when a chemist shows the spatial configuration of somatostatin is such that a particular amino acid is very exposed on the outside of the molecular structure; it may be that by replacing or protecting it, some new activity will be observed."

12. It would be wrong to take differences between what is and is not technical in science as the starting point. These differences are themselves the focus of important negotiations between members. This idea has been especially developed in sociology of techniques by Callon (1975). See also Chapter 1 p. 21ff and Chapter 6.

13. The same tendency is evident in sociological discussions of science which uncritically adopt the attitude that material phenomena are manifestations of conceptual entities.

14. During the first year of the study a new method of chromatography was tried in the laboratory. Albert worked on it for a year trying to adapt it to the purification programme of the group. As soon as it became settled, Albert turned the instrument over to a technician, after which it became a purely "technical" matter.

15. These calculations are only approximate: they are based on the overall budget of the laboratory as computed from grant applications. The activation of the laboratory cost about one million dollars. This was simply to connect the space to the rest of the institute (Photograph 1); buying the equipment on the general market cost approximately $300,000 a year; Ph.D. holders earn an average of $25,000 a year, while for technicians the figure is nearer $19,000 a year. The total wage bill tops half a million dollars a year. The total budget of the laboratory is one and a half million dollars a year.

16. The advantage of a well-kept publication list is that it includes every item produced by the group, including rejected articles, unpublished lectures, abstracts, and so on. The following figures are intended to convey an idea of the scale of article production. Of course, only a stable laboratory can provide a reliable publication list.

17. We use the term "item" to refer to all the different types of published materials, articles, abstracts, lectures, and so on.

18. It is interesting to note the differences between those who argue that the development of a theory of citing behaviour should necessarily precede the use of citation data by sociologists and those who argue that the development of a citation typology will enable the analyst to overcome technical difficulties in the use of citation data. See, for example, Edge (1976) and other contributions to the International Symposium on Quantitative Methods in the History of Science, Berkeley, California, August 25-27, 1976. See also the special issue of *Social Studies of Science* 7 (2; May 1977).

19. In its traditional aristotelian meaning a "modality" is "a proposition in which the predicate is affirmed or denied of the subject with any kind of qualification" (Oxford Dictionary). In a more modern sense, a modality is any statement about another statement (Ducrot and Todorov, 1972). The following discussion owes much to Greimas (1976) and Fabbri (private communication, 1976).

20. The notion of "object" is used here because it has a root in common with "objectivity." Whether a given statement is objective or subjective cannot be determined outside the context of laboratory work. This work is precisely intended to construct an object which can be said to exist beyond any subjectivity (see Chapter 4). As Bachelard (1934) put it "science is not objective, it is projective."

21. In semiotics, the term "deontic" is used to designate the type of modality which indicates what "ought" to be done (Ducrot and Todorov, 1972). Although very crude, this analysis is intended, like the rest of this chapter, to do no more than introduce the general problem of scientific literature. More precise discussion can be found in Gopnik (1973), Greimas (1976) and Bastide (forthcoming).

PHOTOGRAPH FILE

Photograph 1: VIEW FROM THE LABORATORY ROOF

Photograph 2: REFRIGERATOR CONTAINING RACKS OF SAMPLES

93

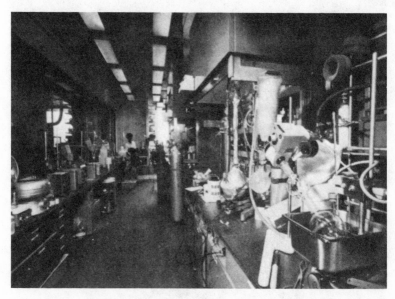

Photograph 3: THE CHEMISTRY SECTION

Photograph 4: A BIOASSAY: THE PREPARATORY STAGE

Photograph 5: A BIOASSAY: AT THE BENCH

Photograph 6: A BIOASSAY: OUTPUT FROM THE GAMMA COUNTER

Photograph 7: FRACTIONATING COLUMNS

Photograph 8: **THE NUCLEAR MAGNETIC RESONANCE SPECTROMETER**

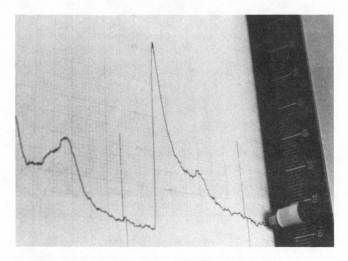

Photograph 9: **TRACES FROM THE AUTOMATIC AMINO ACID ANALYSER**

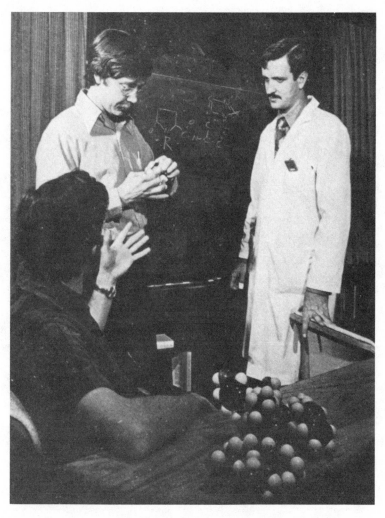

Photograph 10: DISCUSSION IN THE OFFICE SPACE

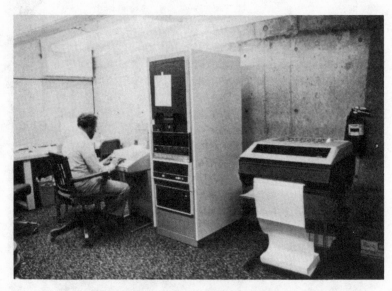

Photograph 11: THE COMPUTER ROOM

Photograph 12: CLEANING UP THE DATA

Photograph 13: AN OFFICE DESK:
                THE JUXTAPOSITION OF LITERATURES

Photograph 14: IN THE SECRETARIAT: TYPING THE FINAL PRODUCT

CT:

In the last chapter, we portrayed an anthropologist making his way through the laboratory and constructing an account in his own terms of the activity he saw. We presented the laboratory as a system of literary inscription, an outcome of which is the occasional conviction of others that something is a fact. Such conviction entails the perception that a fact is something which is simply recorded in an article and that it has neither been socially constructed nor possesses its own history of construction. Understanding the nature of a fact in these terms would obviously hinder any attempt to implement what has been called the "strong programme" in the sociology of science.[1] In this chapter, we shall attempt to examine in detail how a fact takes on a quality which appears to place it beyond the scope of some kinds of sociological and historical explanation. In short, what processes operate to remove the social and historical circumstances on which the construction of a fact depends? In order to pursue this question we confine our discussion to a particular concrete example and to the social construction of a single fact. In particular, we shall specify the precise time and place in the process of fact construction when a statement became transformed into a fact and hence freed from the circumstances of its production.

A fact only becomes such when it loses all temporal qualifications and becomes incorporated into a large body of knowledge drawn upon by others. Consequently, there is an essential difficulty associated with writing the history of a fact: it has, by definition, lost all historical reference. There is a marked difference between a contentious statement and its subsequent (or prior) acceptance as established fact (cf., Chapter 2). Historians of science endeavour to reveal the intervening process of metamorphosis, usually by taking established facts as their starting points and extrapolating backward (for example, Olby, 1974). However, this approach necessarily makes it difficult fully to appreciate a situation in which there is *no* path. Most of the time, historical reconstruction necessarily misses the process of solidification and inversion whereby a statement becomes a fact (see Chapter 4) and this is why some sociologists of science (Collins, 1975) have suggested that it is more useful to monitor contemporary debate than to rely on historical accounts. In spite of these basic methodological difficulties (well known to practitioners of history of science), we shall attempt to reconstruct certain historical events in our laboratory for three main reasons. Firstly, we mentioned in the last chapter that the achievements of the laboratory and the credit bestowed on its members resulted from the characterisation of three substances (TRF, LRF, and somatostatin). The establishment of a new laboratory in 1970 was intended further to develop the achievements of the 1969 programme for the study of TRF. As a result, it was hard to find a single piece of equipment, a grant application, an aspect of behaviour or even a feature of spatial organisation in the laboratory which did not in some way depend on the earlier discovery of TRF. Secondly, the analysis of the construction of TRF turned out to be of manageable size. We were able to accumulate all articles pertaining to TRF (see below for the definition of this corpus), to undertake fifteen interviews with major participants, and to gain access to the archives of the two groups engaged in the TRF(H) research effort.[2] This relatively comprehensive collection of material on one comparatively minor episode provides the basis for a detailed analysis of the social construction of a fact. Thirdly, we have chosen to study the historical genesis of what is now a particularly solid fact. TRF(H) is now an object with a well-defined molecular structure, which at first sight would hardly seem amenable to sociological analysis. If the process of social construction can be demonstrated for a fact of such apparent solidity, we feel this would provide a telling argument for the feasibility of the strong programme in the sociology of science.

In short, our objective in studying the genesis of TRF is simultaneously to provide a background necessary for subsequent chapters, to explicate the influence and the main claims to credit of the laboratory, and to provide support for the view that hard facts are thoroughly understandable in terms of their social construction.

In one sense, historical accounts are necessarily literary fictions (De Certeau, 1973; Greimas, 1976; Foucault, 1966). Historians, as portrayed in historical texts, can move freely in the past, possess knowledge of the future, have the ability to survey settings in which they are not (and never will be) involved, have access to actors' motives, and (rather like god) are all-knowing and all-seeing, able to judge what is good and bad. They can produce histories in which one thing is the "sign" of another and in which disciplines and ideas "burgeon," "mature," or "lie fallow." Our own historical interest in this chapter, however, does not attempt to imitate that of professional historians. We do not attempt to produce a precise chronology of events in the field, nor to determine what "really happened." Nor do we attempt an historical exposition of the development of the speciality of "releasing factors." Instead our concern is to demonstrate how a hard fact can be sociologically deconstructed. With this somewhat lame historical interest we hope to provide an enriched study of the past which avoids some of the basic contradictions and lack of symmetry characteristic of much history of science (Bloor, 1976).

TRF(H) in Its Different Contexts

In order to avoid jeopardising our sociological objective by falling prey to one of the main pitfalls of historical analysis mentioned above, it is important not to start from any knowledge of what TRF(H) "really is." We start, therefore by specifying the way in which the meaning and significance of TRF(H) vary according to the context of their usage.

If we define a network as a set of positions within which an object such as TRF has meaning, it is clear that the facticity of an object is relative only to a particular network or networks. One convenient way roughly to assess the extent of a network is to ask how many people know the meaning of the term TRF (or TRH). We are confident that it would mean little or nothing to the majority of readers. Its expanded form, Thyrotropin Releasing Factor (Hormone), might enable a number of people to connect the term with things scientific. A much smaller group could locate it in endocrinology. To a few thousand medics, for example, TRF refers to a test of use in screening potential malfunctions of the pituitary, although TRH itself would be otherwise no more unusual than other medical substances. To a few thousand endocrinologists, TRH refers to a booming subfield within their discipline. These individuals would recognise TRF as one of a family of recently discovered factors. It is likely that as active researchers these endocrinologists will have read at least some of the 698 published articles (as of 1975) with TRH in the title (see Figure 2.2). If they are physicians, they are likely to have read at least one of the reviews and textbooks which include discussion on the substance. If students, they will have read about TRH in textbooks:

> The most dramatic neuroendocrine discovery made during the period between the present and previous editions of this text was the elucidation of the structure of TRH, accomplished virtually simultaneously by means of investigators associated with the laboratories of Guillemin and Schally (Williams, 1974: 784).

> Some of the hypothalamic releasing and inhibitory factors, which are short peptides have been isolated and identified . . . . They are produced in only minute amounts, for example only 1 mg of the thyrotropin releasing factor (TRF) was obtained from several tons of hypothalamic tissues obtained from the slaughterhouse. Identification and synthesis of some of the releasing and inhibiting factors in the laboratories of R. Guillemin and A.V. Schally and others has been an outstanding advance in biochemical endocrinology (Lehninger, 1975: 810).

In spite of its "outstanding" and "dramatic" character, no more than a few lines are devoted to the discovery in works more than 1000 pages long. For most readers of these texts, knowledge of TRH is limited to these few lines. However, for many researchers and graduate students, TRH is not just a recently discovered structure. It is a substance which can be utilised in setting up new bioassays. To look at, TRH is an unremarkable white powder, which has either been purchased from some large chemical firm or has been given by a colleague. The origin of TRH samples is referred to in articles under sections entitled "Acknowledgments" ("we thank Dr. X for making TRF available to us") or "Materials and Methods" ("TRH was purchased from . . . ."). However, TRH also appears in articles as a well-established fact. Reference to the origin of the concept is made, albeit with decreasing frequency (see Figure 3.1), by means of a tail of perfunctory citations

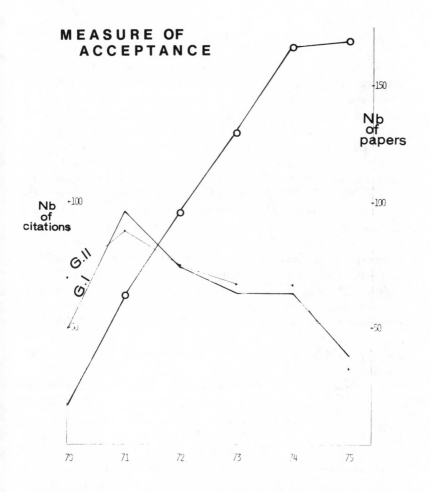

MEASURE OF ACCEPTANCE

Figure 3.1
This diagram combines two sources of information. On the left axis, we plotted the number of citations made to the final articles published on TRF by Schally (G.I) and by Guillemin (G.II). It is clear that the credit—as measured by citations—is nearly identical; it is also clear that less and less citations were made as TRF(H) became a taken-for-granted fact. On the right axis, we plotted the number of *articles* with TRF(H) in the title (see Fig. 2.2). The difference between the slopes of the left and right curves illustrates the transformation of the *fact*.

of one or two of a certain set of eight papers. Within this network, then, TRH is accepted as a fact in the sense that it is sufficient to know that "TRH regulates the release by the pituitary of TSH," that "its chemical formula is Pyro-Glu-His-Pro-NH$_2$" and that it can be purchased from this or that chemical company. At least, this is sufficient to enable the production of articles with titles such as "Investigations concerning TRF induced hypothermia in rats" or "The effect of synthetic TRH on transmembrane potential and membrane resistance of adenohypophysial cells." The main force of arguments in these articles concerns a problem other than the characterisation of TRF; TRF is used simply as a tool. For investigators, such use of a characterised substance, rather than an impure fraction, in an assay means that one of many unknowns can be conveniently eliminated (Chapter 5, p. 205). TRF thus acts as a tool in these papers in the sense that it provides one less concern, or one less source of noise, for the researcher.

For a still smaller group, comprising a few score individuals and half a dozen laboratories, TRH is not merely a tool. For them, TRH represents and entire subfield. Indeed, for a few of the individuals in our study, it represented a lifetime's achievement: TRH constituted their professional life, the justification of their main claim to credit and position.

It is clear, then, that TRF can take on a different meaning and significance depending on the particular network of individuals for which it has relevance. Consequently, a study which focussed on a few individuals within our laboratory is likely to amount to a study of TRF in terms of the careers of these individuals. If a study focussed on wider networks of groups for whom TRF constituted an analytical tool, we would be more likely to emphasise the use of TRF as a technique. Claims about the universality of science should not obscure the fact that TRH exists as a "new recently discovered substance" within the confines of networks of endocrinologists. Its treatment as a nonproblematic substance is confined to a few hundred new investigators. Outside these networks TRH simply does not exist (see Chapter 4). In the hands of outsiders and once devoid of its label, TRH would be merely thought of as "some kind of white powder." It would only become TRH again through its replacement within the network of peptide chemistry where it first originated. Even a well-established fact loses its meaning when divorced from its context.

An additional complication is that networks differ both over space and time, as can be demonstrated by an examination of citations between TRH papers.[3] In 1970 TRH was reconstituted within a new network. Between 1962 and 1970 a group of less than 25 people published 64 papers dealing exclusively with the isolation of TRH, rather than with its modes of action. After 1970, however, TRH appeared in papers published by a much larger group of researchers. The exact interface between the first and the second networks is evident from the continued citing of certain of the pre-1970 papers after the switch took place. Papers dealing with the isolation of TRH were cited 533 times between 1962 and 1970. Between 1970 and 1975, however, they were cited 870 times but almost 80 percent of these citations went to eight papers published between January 1969 and February 1970. The switch from one network to another is also apparent from changes in the authorship of papers with TRH in the title. Before January 1969 almost all authors of TRH papers were neuroendocrinologists engaged in programmes of isolation or in studying modes of action (see Chapter 2). Subsequently, authors came from a variety of neighbouring disciplines. Moreover, there were more outside authors than neuroendocrinologists. These three factors (the number of papers published, the pattern of citations, and the disciplinary origin of authors) indicate the presence of two distinct communities of participants: insiders and outsiders. In addition, we might suppose that the eight highly cited papers provide a clue as to how the meaning of TRH was transformed between one community for whom it represented a life-time's work to another for whom it was merely a technique. How and why this switch took place are the central questions of this chapter.

Even within the network of individuals for whom TRH represented a life-time's work, the precise meaning of the term differed. In the first of the two excerpts from textbooks quoted above it is said that the structure was "accomplished virtually simultaneously by R. Guillemin and A.V. Schally." More strikingly, the second excerpt refers to TRF whereas the first uses the term TRH. We ourselves have used the two terms interchangeably in our discussion so far. In fact, these alternative formulations corresponded directly to those used in each of the groups led by Guillemin and Schally. It became apparent to us that these terms were different names for the same thing by virtue of comments made by members of the laboratory which we studied: what was "actually TRF" was said to have been wrongly referred to elsewhere as TRH. Furthermore, it was argued, credit for the discovery of the substance had been wrongly appropriated by the other group and that what they had identified as a hormone (H) was really a factor (F).[4] Nor did either group agree with the pronouncement that the discovery had been made simultaneously. Instead, each claimed that the other had made the discovery later than themselves and had received credit by virtue of deliberate ambiguities in their accounts of the investigations.[5]

In spite of this controversy between proponents of TRH and TRF, members of a wider network did not heavily favour one or other version. In terms of citations, received credit was evenly divided between the groups, partly because outsiders were not willing to become involved in the dispute, partly because they did not know about it (see Fig. 3.1) and partly because these outsiders were in any case more interested in TRF(H) as a tool than as a contentious scientific achievement. But the mere suggestion that credit had been equally distributed served further to enrage the disputing parties. A member of Schally's group, for example, complained that Guillemin's group had "succeeded in getting half the credit even though they had got there last." A member of Guillemin's group similarly commented that their opponents had obtained half the credit without having done anything. The gradual decrease in citations suggests that it became less and less an issue for the community as a whole as to who actually made the discovery and who should be cited for it. For insiders, however, some bitterness was apparent as much as seven years later. In response to our sociological enquiries (which undoubtedly had the effect of rekindling the dormant conflict) members of each group carefully set out to compare publication and submission dates so as to establish the "correct" and "definitive" allocation of priority.

The Delineation of a Subspecialty:
The Isolation and Characterisation of TRF(H)

We have thus far identified a group of insiders before the end of 1969, and a larger group of outsiders after the end of 1969. The transition from one to another centred around eight papers published in 1969, which were thought to have solved the central research problem. In a similar way, almost all the papers written by insiders before the end of 1969 include references to a few papers published around 1962. Reference to these 1962 papers frequently mention the words "first," "recently shown," "accumulated results," and so on. It is thus possible that developments in 1962 provided a focus for subsequent research in a similar way to the transition that occurred in

1969. On both occasions, a particular cluster of papers provided a starting point. After 1962, a number of papers whose concern was to prove the *existence* of a principle regulating TSH secretion were no longer cited. Instead, references to a smaller cluster of review papers delineated the beginning of new problem. A typical reference to the principle established before 1962, together with a statement of the subsequent problem, is given in the following excerpt:

> Despite accumulated information [9 citations] and *almost universal agreement* that the brain *must* play an important role in the regulation of thyrotropin (TSH) secretion, the nature and extent of this role have *not* been established (Bogdanove, 1962: 622).

None of the nine authors cited in this excerpt participated in the new subspeciality. Prior to the first transition point, research concerned a substance which was universally postulated but the structure of which was unknown. After the second transition point, the nature of the substance was universally accepted but its role and physiological relevance had become problematic. The outcome of research conducted before 1962 could be summarised as "the brain controls TSH secretion." Similarly, the outcome of research conducted before the end of 1969 could be summarised as "TRF(H) is Pyro-Glu-His-Pro-NH_2."

It would of course be possible to delve further into the past in order to determine when and why the inital statement was made about the control of TSH by the brain. For two reasons, however, further retracing would not be useful. Firstly, the statement about TSH was taken after 1962 as a nonproblematic fact and subsequent research into TRF(H) entailed fact production reliant only on the nonproblematic character of the earlier statement. Researchers entering the field of TRF(H) after 1962 could thus rely on Bogdanove's (1962) review as sufficient baseline information. Secondly, in order to achieve an understanding of the construction of facts, it is necessary to focus on one specific episode rather than on longer periods. The study of longer periods would necessitate our accepting a larger number of facts without examining their construction.

We constructed a file of all articles published between 1962 and 1969 which deal exclusively with the isolation of TRF(H). This file was initially built up from lists of articles in the two laboratories which worked on TRF(H) and from citations in these articles. The file was checked against *Index Medicus* and double-checked with the *Science Citation Index* (SCI) *Permuterm,* as a result of which review articles were added. In total, four groups have worked on the isolation of TRF: Two groups, led by Schibuzawa in Japan and by Schreiber in Hungary, left the field after a while for reasons which will later become clear; Schally's group took up TRF(H) research in 1963; only Guillemin's group maintained research activity throughout the period 1962-1969. A few other authors wrote reviews but did not participate in the citation network (in other words, although they cited other papers, they were not themselves cited). Articles dealing with mode of action of TRF(H), rather than its isolation, were excluded.

Figure 3.2 is a schematic representation of the growth of TRF(H) subspecialty between 1962 and 1969 (inclusive). The vertical axis represents time and the horizontal axis represents the cumulative number of papers cited by TRF(H) papers. Thus, each published paper is plotted on the diagram according to (a) its date of publication (b) the number of new citations it made over and above those previously made by earlier papers. For a specialty whose papers continually cited the same body of material, we would expect a more vertical growth curve. Two features of growth are apparent from the curve for TRF(H). Firstly, there was an increased rate of publication during two phases of development; during 1965 and during 1969. Secondly, at several points, papers were published which drew heavily upon a fresh body of cited material. These points, represented on the curve by "kicks" to the left (and labelled with arrows in the diagram), occurred in 1962, 1965, 1966, and 1968. As we shall see later, the shape of this curve matches the recollections of informants as expressed through interviews. For example, the sudden increase in newly cited material in 1966 corresponds to the entry into the field by Schally's group. By contrast, the sections of the curve which are almost vertical correspond to what respondents referred to as depressive and nonproductive periods.

A Choice of Strategies

Clearly, there are disadvantages in relying solely on the description of an area delineated in terms of publications and citations. In particular, it is all too easy when thinking of a research area in these terms to think of its boundaries as objectively independent of participants. To counter this effect we shall use additional material to show how the area may well have developed in a different direction.

MEASURE OF GROWTH

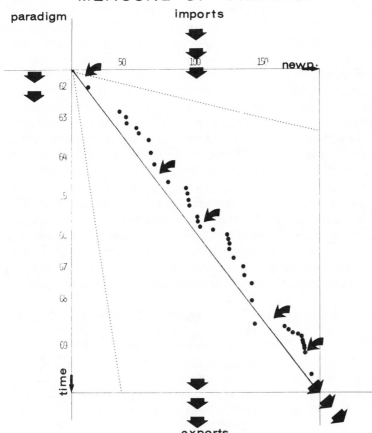

Figure 3.2
This is a schematic representation of the growth of the TRF specialty.
Each point represents one paper; the horizontal axis represents the
cumulative number of papers cited by these papers; the vertical axis
represents time. The upper limit (top left hand corner) corresponds to
the end of the controversy about the *existence* of TRF; the lower limit
(bottom right hand corner) correspondes to the end of the controversy about *what* TRF is. The number of papers, the distance of each
paper from the preceding one—both in terms of time and number of
newly cited material—provides a general pattern which is strikingly
different between one area and another. The general shape of the
diagram illustrates the importance of imported papers (see Fig. 3.4)
and of citing papers from other areas. In this representation, each
paper is related to all the papers it cites and all the papers it is cited by.
The general map—impossible to retrace here—gives an approximation of the field and of all the operations which are performed in it.
115

By 1962, a number of hormones other than TRF(H) had been discovered (Meites et al., 1975; Donovan et al., forthcoming). Indeed, after the war endocrinology had been completely transformed by the determination of the amino acid constituents and sequences of several hormones (such as insulin, oxytocin and vasopressin). Thus, the anticipation that a sequence could be found for TRF was not new. However, the pursuit of this sequence entailed difficult and risky decisions. In order to appreciate that the TRF(H) research programme was based on decisions about an uncertain future, rather than on logical deductions from past events, it is necessary to examine the alternative courses of action which were then possible. Firstly, no other hypothalamic factors had been characterized by 1962. The analogy with hormones successfully discovered remained very much an analogy, as the use of the term factor indicates (Harris, 1972). Although physiological investigations of hypothalamic factors were making solid advances, almost no progress had been made in chemical investigations. According to the majority of participants, the number of unsubstantiated claims prevalent at this time was overwhelming. Their frustration was made explicit in many papers of the period:

> The young field of hypothalamic pituitary physiology is already littered with dead and dying hypotheses. I probably add to the casualties by presenting another premature proposal (Bogdanove, 1962: 626).
>
> The oddity of the situation with respect to the hypothalamic substances is that never before to my knowledge, except for the monster of Loch Ness and the Abominable Snowman of the Himalayas has the existence of hypothetical objects been indicated by so much imposing circumstantial evidence (Greep, 1963: 511).

An eminent pharmacologist commented similarly: "the only thing I can believe in this field are the retractions" (Guillemin, 1975). By 1962, work on the first postulated factor (CRF, see Chapter 2) was at the same stage it had been for the previous ten years and was to reamin thus for the next fifteen years. A host of factors had been postulated which remained unconfirmed in 1976 and artefacts abounded (Chapter 4). Virtually any consistent effect was given a name and a few preliminary steps of purification from the soup of brain extracts were considered sufficient to merit the writing of a paper. Frequently, the effect was regarded as sufficiently consistent to merit writing a paper on aspects of rat behaviour, calcium levels, or thermoregulation.

Secondly, the decision to begin research on TRF(H) entailed the postulation both of the existence of *new* discrete factors and that these factors were peptides. Although, at the time, the notion that the brain regulated the pituitary was a prerequisite to being a neuroendocrinologist, it was also possible to hold that *known* factors like oxytocin and vasopressin could account for such regulation. For instance, as late as 1969, one of Guillemin's papers was rejected by *Science* simply because "it was well known that vasopressin releases TSH *in vitro* and *in vivo.*" Another investigator, McCann, was not interested in TRF, which he regarded as an artefact, the effect of which could be explained by recourse to a known substance (Donovan et al., in press). Sticking to the idea that there was a new factor entailed a further assumption that the factor was a peptide, because this was the only way that available chemistry could be utilised in the releasing factor field. So the postulation was dual. The substance had to be new, but the chemistry of this new object had to be classical and was imported from outside fields after appropriate modifications. We shall return to this point later.

Thirdly, the strategy of isolating and characterizing substances, although already well established by Du Vigneaud's achievement with vasopressin and oxytocin, was slightly at odds with the physiological training of neuroendocrinologists. For example, although Harris, Scharrer, McCann, and Guillemin were all expert at setting up sophisticated bioassays, at growing cell cultures, and at preparing anatomical slides, they were largely ignorant of chemistry. Chemistry was "ancilla physiologicae" to them. If Harris and McCann accepted the idea of undertaking some isolation work, they never accepted the relegation of physiology to a discipline subservient to the goals and practices of chemists (Harris, 1972). One of their arguments concerned their distaste for teaching duties and the extreme dullness of routine chemistry.

> When you have students you cannot ask them to cut brains all the time: you have to give them interesting things to do: you cannot corner them in routine tasks, which will pay off only in five or six years. If they come to your laboratory in order to graduate, they expect to write a few papers, it has to be interesting (McCann, 1976).

The decision to obtain the structure of TRF(H) also entailed considerable expense, because if these peptides existed at all, they could only be found in minute quantities (thousands of times smaller than the hormones which had been characterized by Du Vigneaud). The gathering and treatment of millions of hypothalami was a collossal task. As Schally put it:

> People became suspicious . . . they were used to high yield peptides like the others [oxytocin] . . . and they could not understand why we didn't get the structure . . . It was not fair of them, we had to create a whole technology . . . .Nobody before had to process millions of hypothalami . . . The key factor is not the money, it's the will . . . the brutal force of putting in 60 hours a week for a year to get one million fragments (Schally, 1976).

An idea of the resistance to this programme can be gained by comparing the strategy adopted by Guillemin with that of Harris, who was one of the founders of the field. Even after he had hired one chemist whose sole task was to isolate LRF, Harris maintained a slow and cumbersome assay on conscious rabbits which prevented the chemist from screening more than five or eight fractions a month. If the chemist had been allowed to work at his natural speed, he would have produced far more fractions than the physiologist could possibly have coped with. Usually, however, the chemist had to give way; the physiologist maintained his assay which he perceived as more interesting. Of course, as one of Harris's former colleagues commented:

> [H]e wanted that isolation done . . . but he didn't devote very much of his overall effort to help isolating these factors . . . being basically a neuroanatomist. . . . I was able to convince him to have hypothalami shipped from the US . . . *we went that far* . . . he could not guess that we would need 100 times that amount (Anonymous, 1976a).

Schally's strategy was completely different:

> I am not interested in Physiology. . . . I want to help physicians, clinicians . . . and the only way is to extract these compounds, isolate them and provide clinicians with enormous quantities of them . . . like vitamin C. Somebody had to have the guts . . . now we have tons of it. . . .
>
> That's why I chose extraction . . . there is no choice. It's like fighting Hitler. You have to cut him down. It's not a choice. The strategy was good and was the only one (Schally, 1976).

The decision to redefine the TRF subspecialty solely in terms of determining the *structure* of the substance completely reshaped the professional practice of the subfield, even though this was entirely in line with the central concepts of endocrinology as a whole. Precisely because of the consistency of Guillemin's strategy with the objectives of endocrinology, his decision did not constitute an intellectual revolution.

Because of the success of his strategy, there is a tendency to think of Guillemin's decision as having been the only correct one to make. But the decision to reshape the field was not logically necessary. Even if the decision to pursue the structure of TRF(H) had not been taken, a subfield of releasing factors would still exist. Of course, only crude or partially purified extracts in short supply would be used, but all the problems of physiology could nonetheless be studied, if not resolved. It should also be realised that until 1969 there was no indication that the strategies adopted by Guillemin and Schally were successful. Indeed, everything that occurred before 1969 suggested that it was folly to reshape the specialty in 1962. Similarly, it was thought that Guillemin would have been better off to wait for a drastic amelioration in peptide analysis which would have then made possible the solution of the TRF problem by the use of picogram quantities at a much lower cost (Anonymous, 1976b).

The Elimination of Concurrent Efforts by New Investments

It is probably no coincidence that the two researchers (Guillemin and Schally) who dared to plunge themselves into the task of reshaping the field were both immigrants. Schally's testimony is particularly suggestive of the importance of their initially peripheral position. For example, he made the following remarks about a third party:

> He is the Establishment . . . he never had to do anything . . . everything was given to him . . . of course, he missed the boat, he never dared putting in what was required: brute force. Guillemin and I, we are immigrants, obscure little doctors, we fought our way to the top; that's what I like about Guillemin; at least we fought, and [with a gesture of his hand towards framed awards on the wall] now we have more awards than all of them (Schally, 1976).

The present case appears to fit fairly well with what is known about the formation of specialties. The enormity of the research task tended only to attract people who were not in a position to satisfy themselves with physiology, and who were not prepared for a conceptual revolution. They occupied a niche which entailed a break with existing methods and an immense amount of hard, dull, costly, and repetitive work: the kind of niche from which people normally shun away.

The enormity of the research task and the nature of the decision explain why more people did not take up this work. This is also consistent with the fate of investigators who dropped the question after making some initial contributions. One reviewer, for example, drew attention to the "misguided work" done by Schibuzawa and Schreiber as follows:

> Schibuzawa and his colleagues have been studying a polypeptide which they can extract from the hypothalamus and posterior pituitary lobe . . . . They call it TRF (thyrotropin releasing factor) and believe it to be the neurohumor. . . . Their findings have not been confirmed so far (Bogdanove, 1962: 623).

Schibuzawa apparently made the same choices as Guillemin. He claimed to have isolated TRF and even presented an amino acid composition for his peptide. But instead of being acclaimed as having solved the problem of TRF in two years, his work was surrounded by questions. His papers were criticised word by word and his fractions were said not to display any activity in laboratories other than his own. According to one account, he did not turn up when invited to repeat his experiments at one laboratory. In terms of our discussion in Chapter 2, the operations on his papers took the form of doubt and deprecation. He wrote no new papers after 1962, his claim to have solved the TRF problem was dissolved, and his substance became regarded as an artefact. Subsequently, he left research altogether. It is important to note that despite Schibuzawa's inability to prove his claims at the time, they were proved (with the exception of the amino acid composition) ten years later. This was not so much because of his failure but rather because in the meantime the definition of a proof had drastically changed.

Schibuzawa's claims were unacceptable because somebody else entered the field, redefined the subspecialty in terms of a new set of rules, had decided to obtain the structure at all costs, and had been prepared to devote the energy of "a steam roller" to its solution. For Schibuzawa it had had been sufficient to draw upon the existing stock of accumulated knowledge and to touch upon questions of isolation while essentially remaining within classical physiology.

> This was what you could call "normal science" . . . Thus, everybody knowing the field could make deductions as to what TRF was . . . their conclusions were correct, but it took ten years to prove it. . . . To this day, I don't believe they had ever seen what they talked about. They, Schibuzawa and Schreiber both wrote one too many papers giving the amino acid compositions. Now here, there is no logical assumption. There is no way you can postulate the amino acid composition of an unknown substance (Guillemin, 1975).

In other words, there was no easy shortcut between what was already known and the problem of sequence. Since Guillemin wanted to determine the *sequence* of TRF, and since he was ready to reshape the subfield around this crucial goal, new standards were set as to what could and could not be judged reliable. Data, assays, methods, and claims which might have been acceptable in relation to other goals, were no longer accepted. Whereas Schibuzawa's papers might previously have been accepted as valid, they were subsequently regarded as wrong. That is to say the epistemological qualities of validity or wrongness cannot be separated from sociological notions of decision-making.

The sudden change in the criteria of acceptability was made explicit in a long review paper published in French (Guillemin, 1963). This review specified fourteen criteria which had to be met before the existence of a new releasing factor could be accepted. These criteria were so stringent that only a few signals could be distinguished from the background noise. This, in turn, meant that most previous releasing factor literature had to be dismissed (Latour and Fabbri, 1977).

> These rigorous criteria contribute to take away any meaning from a great number of publications which hastily concluded that this or that substance acts only through the stimulation of the secretion of a pituitary hormone, or even that this or that protocol fits this explanation alone (Guillemin, 1963: 14).

In one important sense, then, TRF did not exist prior to the imposition of limitations, because such limitations preceded the first experiments and defined what could be accepted in advance. In his paper, Guillemin argued that prior to that time, the field had been characterised by artefacts, unfounded claims, and elegant hypotheses rather than by facts. On the basis of this reconstitution of the past in terms of artefacts, Guillemin proposed criteria each of which was designed to eliminate a priori any future possibility of an artefact, or, at least, any possiblity of an artefact within the new context.

The acceptance of these criteria demanded the expense of investing in equipment which would meet the necessary stringency. Consequently, each of the criteria specified in the review article was responsible for the import into the laboratory of items of equipment necessary for constructing TRF.

> The *physiological validation* of a substance of hypothalamic origin as being a hypophysiotropic mediator, is thus a considerable enterprise; it requires multiple and sometimes complex techniques in neurophysiology . . . in biochemistry to fulfil all the above conditions before asserting that this hypothamalic substance or fraction, is a hypophysiotropic mediator (Guillemin, 1963: 14).

The same source also points out the difficulty of meeting the criteria and the cost of the corresponding investment.

> Such a project can be undertaken only by a group, a team in which everyone has different but complementary skills for the central idea around which the team has been conceived and realized. This is certainly the necessary characteristic of this new orientation of Physiology, that is Neuroendocrinology (Guillemin, 1963: 11).

The consequence of this new investment was immediately reflected by Harris's strategy. The rules of the game, as defined by Guillemin, became so stringent that one of Harris's chemists gave up this line of research

> . . . [B]ecause I knew what we were competing against in this country [USA] in terms of money, scale of work . . . and there were no ways we could achieve parity, if you like, in England at that time (Anonymous, 1976a).

The requirements imposed by the new strategy were noted in subsequent articles which contain assessments of work by Schibuzawa or Schreiber. These assessments consisted largely of qualifications which had the effect of discrediting earlier contributions. Such phrases as "gratuitous affirmation," "assays not specific enough anymore," "not really demonstrated," and "unreliable" are common. By contrast, Guillemin's group's first (1962) paper was widely acclaimed (for example, it was said to be the "first uncontrovertible evidence") and was similarly received in subsequent years. None of the 90 citations to this paper (as listed in the SCI between 1963 and 1969) were negative (Latour, 1976).

The results of the fresh accumulation of constraints was to put Schreiber out of the race. By increasing the material and intellectual requirements, the number of competitors was reduced. According to one of his colleagues, Schreiber withdrew for various material and strategical reasons.

> His acid phosphatase test was not really good: it was heavily criticised . . . he was wrong on the amino acid composition . . . he had coherent ideas about the subject and was running proper experiments but at that time it was very difficult to get hypothalami . . . he had to do it himself; no one realised that you need not 200 of them, but 20,000 of them . . . he then realised he simply could not compete . . . also you could not obtain radioactive iodine of high specificity, we had to wait half a year to have it from England, so he could not do the assays . . . it does not make sense to spend time on a field when you cannot compete (Anonymous, 1976b).

The same informant also provided a more ideological explanation for Schreiber's withdrawal:

> After the communist takeover in Prague, Endocrinology was not well in favour . . . at that time the connection between the nervous system and endocrine system was not very clear—the feedback theory, at this time triumphant, they did not accept because it was a self contained system . . . that is the reason why I did not go into Endocrinology . . . the whole milieu was antagonistic to endocrinological research . . . There was a span of 5 or 7 years before we could work again, and not only conditioned reflexes (Anonymous, 1976b).

This provides an example of the perceived influence of macrosociological factors on the field, rather than that of multiple fine social determinations with which we have so far been mainly concerned. It is worth noting, however, that this statement provoked dismissive comments from other participants. Guillemin, for example, felt that such statements of ideological influence were mere rationalisations of the real fact that Schreiber had "missed the boat."

The decision drastically to change the rules of the subfield appears to have involved the kind of asceticism associated with strategies of not spending a penny before accumulating a million. There was this kind of asceticism in the decisions to resist simplifying the research question, to accumulate a new technology, to start bioassays from scratch, and firmly to reject any previous claims. In the main, the constraints on what was acceptable were determined by the imperatives of the research goals, that is, obtain the structure *at any cost*. Previously, it had been possible to embark on physiological research with a semipurified fraction because the research objective was to obtain the physiological effect. When attempting to determine the structure, however, researchers needed absolutely to rely on the accuracy of their bioassays.

The new constraints on work were thus defined both by the new research goal and by the means through which structures could be determined. As a result of these constraints, we have seen that researchers such as Schibuzawa, Schreiber, and Harris were excluded. But for the support of funding agencies, Guillemin might himself have remained a mere critic of others' work. But Guillemin's past achievements provided some guarantee that he could carry out research on the basis of the new constraints.[6] Even so, no one expected in 1962 that the determination of structure would take eight years, millions of hypothalami and more asceticism than anyone could have guessed.

The Construction of a New Object

We started by identifying the different networks in which TRF had meaning and by surveying the area in which it was created. We then discussed how a point of transition opened up the TRF area and how a new research imperative, "obtain the structure at all costs," subordinated the role of physiology relative to chemistry. This new strategy had the effect of both raising the cost of the programme and increasing the stringency of the rules. It was acknowledged as commendable by neuroendocrinologists as a whole and funded by U.S. agencies. However, the new strategy effectively eliminated competition from Japan, Czechoslovakia, and England. We can now turn our attention to the TRF area itself.

Guillemin's initial decision was to determine the structure of any one releasing factor. The specific choice of TRF, was in fact, due to a variety of reasons. After the long failure of work on CRF, Guillemin's group became interested in LRF because of McCann's new assay. Guillemin also decided to set up a new assay based on the principle of the McKenzie assay, a classical test for measuring TSH, because a technician who had previously worked on TSH, came to the laboratory.

> I was not sure how much of Schibuzawa and Schreiber should be accepted, so I didn't want to put too much of my time into TRF . . . within six months the assay worked relatively well (Guillemin, 1976).

At first these research efforts constituted a secondary programme: "Then, it became obvious to me that we could look at TRF" (Guillemin, 1976). However, this was not done to check Schreiber's claims.

> No, I neglected those, and it was not to check them, if you start checking that sort of thing you never do anything; the idea was to start a completely *de novo* bioassay for TRF (Guillemin, 1976).

But this kind of assay was widely available at that time:

> To this day I don't understand how Schreiber could use this ludicrous assay, while anybody could have done what we did in 1961 and set up a true assay for TRF . . . it was simple, everything was available . . . classical endocrinology (Guillemin, 1976)

A new object of study thus came about within a period of normal science as a result of the facilities of classical endocrinology, together with the benefit of one technician's expertise and the raising of requirements due to Guillemin's strategical decision. The new object took its initial existence within a local context but soon attracted a great deal of outside attention. However, it is crucial not to use hindsight to define this new object; it was *not* the TRF of 1963, 1966, 1969, or 1975. From a strictly ethnographic point of view, the object initially comprised the *superimposition* of two peaks after several trials. In other words, the object was constructed out of the *difference* between peaks on two curves. Let us try to clarify this point by outlining the process whereby a new object begins to be constructed.

Initially, a curve produced by a bioassay is taken as a baseline against which variations can be contrasted. Subsequently, an "elution curve" is produced by a bioassay on a purified fraction (see Chapter 2). After each purified fraction has been tested for bioactivity, the two curves are superimposed. If there is a discernible discrepancy between the control curve and that for the purified fraction, the fraction may then be referred to as a "fraction with TRF-like activity." As we have seen, however, these kinds of claims about the presence of substances and activities are common. Frequently, discrepancies between curves are subsequently shown to have arisen as a result of background noise in the bioassays, at which point the bioassays are denounced as insufficiently stable and the claim to have found a fraction is dissolved. However, when the same fraction is seen to give rise to the same activity, the initial claim begins to be taken more seriously. In other words, criteria of repetition and similarity are sufficient to begin to *substantiate* the initial claim. Consequently, the fraction is referred to as an entity with some consistent qualities and the initial label (TRF) begins to stick. Even so, practitioners are wary of categorical statements that the substance actually is TRF.

The steady activity constituted by repeated bioassays might have been caused by a well-known substance such as oxytocin. The application of the constraints outlined above then enables a distinction to be made between the new substance and any other known activity. In brief, these constraints require a signal unlike any other expected signal to be discerned from the background noise. If such a distinction is identified, the substance is taken as both stable, distinct, and new.

Despite the fact that this process was not novel, its use in Guillemin's laboratory resulted in a new object (fraction with TRF-like activity) which neither dissolved between one trial and another, nor between one purification step and another. In addition, this object (unlike Schibuzawa's and Schreiber's fractions) did not become the focus of controversy. The multiple precautions taken through statistical analysis, the reputation of the laboratory, and use of assays (for MSH, oxytocin, vasopressin, LRF, CRF, and ACTH) all countered any possible objections which could be made by colleagues.

Although the repeated overlapping of two peaks in 1962 was taken to indicate the presence of a new discrete entity, it was not claimed that they had found a substance. This was because the amino acid composition and sequence of the entity had not then been obtained. It was still possible that a corresponding substance might never be obtained, as had been the case with CRF. Even if a sequence was subsequently found, the substance could still turn out to be an artefact, as might still be the case with TRF itself (see Chapter 4). We thus need to stress the importance of not "reifying" the process by which a substance is constructed. An object can be said to exist solely in terms of the difference between two inscriptions. In other words, an object is simply a signal distinct from the background of the field and the noise of the instruments. Most importantly, the extraction of the signal and the recognition of its distinctiveness themselves depended on the costly and cumbersome procedure for obtaining a steady baseline. This, in turn, was made possible by laboratory routine and by the iron hand of the scientist who organised laboratory work and took all precautions available within the laboratory context. Once again, to say that TRF is constructed is not to deny its solidity as a fact. Rather, it is to emphasise how, where, and why it was created.

The list of technical papers published by Guillemin's group between 1962 and 1966 gives an indication of the context within which TRF was constituted as a stable object.[7] Firstly, the majority of technical citations were made by TRF articles to other TRF articles. This indicates the internal response of the subspecialty to the new set of constraints imposed by Guillemin's strategy. Secondly, articles published in the first years of the subspecialty were predominantly cited. These early papers thus appear to have formed the technical basis of future operations. Thirdly, several techniques were borrowed from other projects current in the group (for example, assays for LRF and CRF). Fourthly, a number of techniques were imported from neighbouring fields. This external borrowing occurred at crucial points in the development of the TRF field. Citations are made to techniques, statistics, and enzymology in 1962; and mostly to biochemistry in 1966 and 1968. We can thus see, on the one hand, that the construction of TRF depended on author's provision of inscriptions obtained from instruments accumulated within the laboratory. At the same time, the *solidity* of this object, which prevented it from becoming subjective or artefactual, was constituted by the steady accumulation of techniques.

Before 1966, TRF articles were primarily concerned with the arrangement of instruments and with improving purification processes. These predominantly technical concerns necessarily presupposd the existence of TRF and hence made possible the further purification of a fraction. By 1966, an almost pure material had been obtained, which was then subjected to the analytical tools of Chemistry. (Although an amino acid composition of the material had already been obtained in 1965, it was not then generally regarded as correct.) However, after this rapid advance, the programme was pulled up by an unexpected practical problem:

Perhaps the most obvious comment suggested by the results reported
here relates to the large number of brain fragments (hypothalamus) that
were necessary for purification of a small quantity of the hypothalamic
neurohumor. Obviously, a much larger number of brains will be
necessary to provide enough of the polypeptide to approach its amino
acid sequence... Thus the problem of availability of large quantities of
hypothalamic fragments collected in adequate conditions . . . remains
the absolute prerequisite for a meaningful programme on isolation
(Guillemin et al., 1965: 1136).

This situation was particular to the field of releasing factors. For
endocrinology as a whole, sufficient quantities of hormones had
always been available. However, attempts to obtain the structure of
releasing factors were constrained by difficulties in obtaining sufficient
quantities of hypothalamus.

From the perspective of 1966, it remained perfectly possible that
the programme would be subsequently phased out. It was then feasible
that partially purified fractions would be continued to be used in the
study of modes of action, that localization and classical physiology
could have continued, and that Guillemin would merely have lost a few
years in working up a blind alley (Anonymous, 1976b). TRF would
have attained a status similar to GRF or CRF, each of which refers to
some activity in the bioassay, the precise chemical structure of which
had not yet been constructed.

One important feature of our discussion so far is worth noting at this
point. We have attempted to avoid using terms which would change the
nature of the issues under discussion. Thus, in emphasising the process
whereby substances are *constructed*, we have tried to avoid descriptions of the bioassays which take as unproblematic relationships
between signs and things signified. Despite the fact that our scientists
held the belief that the inscriptions could be representations or
indicators of some entity with an independent existence "out there,"
we have argued that such entities were constituted solely through the
use of these inscriptions. It is not simply that differences between
curves indicate the presence of a substance; rather the substance is
identical with perceived differences between curves. In order to stress
this point, we have eschewed the use of expressions such as "the
substance was discovered by using a bioassay" or "the object was
found as a result of identifying differences between two peaks." To
employ such expressions would be to convey the misleading impression
that the presence of certain objects was a pregiven and that such

objects merely awaited the timely revelation of their existence by scientists. By contrast, we do not conceive of scientists using various strategies as pulling back the curtain on pregiven, but hitherto concealed, truths. Rather, objects (in this case, substances) are constituted through the artful creativity of scientists. Interestingly, attempts to avoid the use of terminology which implies the preexistence of objects subsequently revealed by scientists has led us into certain stylistic difficulties. This, we suggest, is precisely because of the prevalence of a certain form of discourse in descriptions of scientific process. We have therefore found it extremely difficult to formulate descriptions of scientific activity which do *not* yield to the misleading impression that science is about *discovery* (rather than creativity and construction). It is not just that a change of emphasis is required; rather, the formulations which characterise historical descriptions of scientific practice require exorcism before the nature of this practice can be best understood.[8]

The Peptidic Nature of TRF

1966 marked the end of a period of hard but successful work and the beginning of three years' frustration. A basic assumption which had thus far guided the choice of procedures and the use of analytical tools was that TRF was a peptide. This assumption was taken as noncontroversial at the initial stages of the specialty. However, the peptidic nature of the substance was a contextual definition. In particular, this definition could be reaffirmed by the resistance of a fraction to a long series of trials involving the use of several enzymes. The substance was deemed to be a peptide if its activity was destroyed during the course of these trials. For example, one paper written in 1963 had confirmed the peptidic nature of the material after a first set of such trials:

> In this note we show arguments in favour of the peptidic nature of these substances; their biological activity is destroyed partially or totally, by pepsic or trypsic digestion and by heating in presence of hydrochloric acid (Jutisz et al., 1966: 235).

In addition, past experience had led participants to anticipate an increase in the ratio of amino acid as a purer and purer fraction of the peptide was obtained. In 1964, however, this increase had failed to materialize. Moreover, a new set of enzymatic tests had failed to destroy the activity of the fractions. The conclusiveness of the tests depended both on the number of enzymes used and on how well their action was characterized. By 1966, the list of enzymes used in the test had grown extensive but none could destroy activity in the required way. It was logical to conclude that the substance was not a peptide. In fact, one enzyme added to the list a few years later did succeed in destroying the activity of the fraction. By this time, however, the substance had already been "proved" to be a peptide. This thus demonstrates that proof and the reaching of logical conclusions depend entirely on context, in this case on the availability of certain enzymes.

In papers published in May 1966, Guillemin's group drew the logical conclusion from the negative results:

> These results are compatible with the hypothesis that TRF might not be a simple polypeptide as hitherto thought (Burgus et al., 1966: 2645).

> We have been led to question the long held hypothesis that TRF and LRF are of peptidic nature (Guillemin et al., 1966: 2279).

Participants found only an extremely small percentage of amino acids in their purest sample. The possibility thus arose that a large component of TRF was of a completely different chemical nature. It followed that the appropriate equipment and procedures for its study might be different. The meaning of TRF thus changed. Consequently, it was likely that the chemistry which was borrowed to study the substance would be modified and that there would be some significant effects on the organisation of the specialty.

The new hypothesis, that TRF comprised a small peptidic component and a large nonpeptidic component, was confirmed as a result of work by Schally, who was a newcomer to this problem. Schally had been a former postdoctoral fellow in Guillemin's laboratory. He provided a vivid contrast to Guillemin's cautious, positivist approach. Whereas Guillemin had talked mainly in terms of methods, Schally talked about strategy. He portrayed his attempts to collect vast quantities of hypothalami in terms of his use of "guts and brute force." He claimed that Napoleon's campaigns provided the inspiration for his scientific method, and he talked of the TRF specialty in terms of a "battle field" strewn with the corpses of competitors. "He is a dynamo," commented another participant. He was able directly to supervise the purification part of the TRF programme because of his training in chemistry, and relied on a physiologist for the operation of the bioassay. By contrast, Guillemin was a physiologist by training who had to rely on someone else for precise chemical work. Neither of them liked having to rely completely on someone else's expertise, but this necessity was dictated by their perception of the problem.

By the time Schally published on TRF in 1966, Schreiber had already withdrawn and Guillemin's group was alone in the field. The methodology taken up by Schally was essentially the same as Guillemin's except that they respectively worked with porcine and ovine brain extracts. But despite the fact that each of the two groups led by Guillemin and Schally worked in the same area and used similar methods, there was an essential difference in their beliefs.[9] In particular, Guillemin's group did not believe that results produced by Schally's group to the same extent that Schally's group believed those produced by Guillemin. This asymmetry helps explain why Schally went on to confirm the nonpeptidic nature of TRF.

Between 1962 and 1969 (inclusive), the two groups published a total of forty-one papers concerned exclusively with the isolation and characterisation of TRF. Of these, twenty-four were published by Guillemin's group and seventeen by Schally's group. This difference in output reflects the fact that TRF was the main programme of Guillemin's group for eight years, whereas it was only a secondary programme for Schally's group for four years. As late as 1969, Schally remarked that he was not interested in TRH.

The pattern of citations also reveals a marked asymmetry between the two groups. Whereas Guillemin's group cited their own articles in the TRF area one hundred and three times, they cited Schally's papers on the same question a mere twenty-five times. Schally's group, on the other hand, cited their own papers (forty-seven times) with roughly the same frequency as articles by Guillemin's group (thirty-nine times). Whereas Guillemin's group cited papers within the group but outside the TRF field only twenty-eight times, Schally's group cited the equivalent body of papers fifty-seven times. This tends to indicate that Guillemin's groups had constituted a new methodology on which they relied heavily, whereas Schally's group was more reliant both on Guillemin's work and other outside sources.

If we now consider the nature of citations between the groups, rather than just their number, the asymmetry is even more striking.[10] For all citations of Guillemin's papers by Schally (and vice versa) we identified the nature of citing operations in terms of borrowing or transforming. Figures 3.3a and 3.3b represent citations by Guillemin

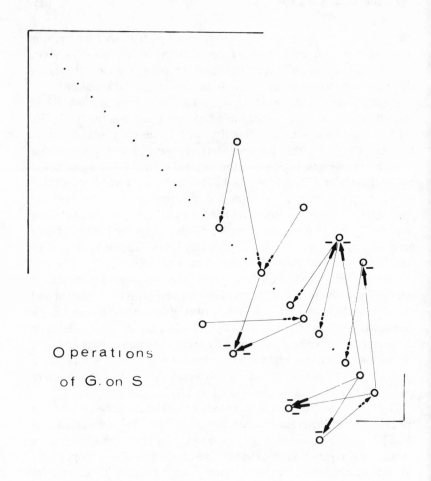

Operations

of G. on S

Figure 3.3a & b
This figure is derived from Fig. 3.2. Only the major publications are shown, and those of Guillemin's (G) and Schally's (S) groups are separated for illustrative purposes. In both figures 3a and 3b, Guillemin's group's papers lie along the diagonal, with those of Schally's group on either side of the diagonal. The main citing operations of each group on the other's papers are represented in simplified form by arrows between papers. *Borrowing* operations are represented by arrows from cited to citing papers; *transforming* operations are represented by arrows from citing to cited papers. The plus and minus signs indicate the sense of transformation.

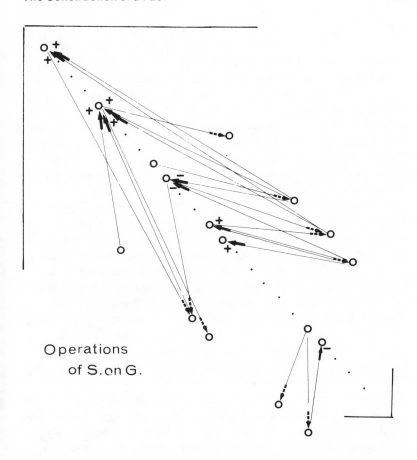

Operations
of S. on G.

of Schally and Schally and Guillemin respectively. In these figures, *borrowing* operations are represented by arrows from cit*ed* to cit*ing* papers. For *transforming* operations, the arrow is in the reverse direction. In addition, a plus or minus sign indicates whether a transforming operation was a *confirmation* or *refutation*. The figures show that all citations made by Schally were directed towards Guillemin's initial work and constituted either borrowing or confirming operations (apart from two negative citations of one paper). This reflects the fact that Schally found it unnecessary to modify Guillemin's findings. By contrast, almost all the citations made by Guillemin constitute negative transformations. Closer inspection reveals that those of Guillemin's citations that performed borrowing operations were made to papers by Schally which had previously confirmed work by Guillemin. For example, one of Guillemin's papers contains the comment that "this paper [reference to a paper by Schally's group] confirmed our former hypothesis." Such differences are too striking to be interpreted simply as differences in citation practice. Instead, we propose that they reflect an essential asymmetry of confidence between the two groups.

We have already suggested that the meaning of TRF(H) was negotiated with reference to particular contexts which comprised both the material layout of the laboratories and the particular strategies adopted by the two competing groups. This is best illustrated by way of an example.

In 1966, Schally published a paper in the wake of Guillemin's suggestion that TRF might not be a polypeptide. The tentative suggestion earlier put forward by Guillemin's group—"These results are compatible with the hypothesis that TRF may not be a simple polypeptide" (Burgus et al., 1966)—was borrowed as a quasi-fact in Schally's 1966 paper: "purified materials appear not to be a simple polypeptide since amino acids account for only 30% of its composition" (Schally et al., 1968). As we have already noted, a low concentration of amino acid could be taken as establishing either that the substance was not pure or that it was not a peptide, according to context. Schally's belief in Guillemin's new hypothesis persuaded Schally to accept the interpretation that TRF(H) was not a peptide. This would be unremarkable but for the fact that in accepting this interpretation, Schally was invalidating the amino acid composition which he himself found: "After hydrolysis TRF was shown to contain 3 amino acids, histidine, glutamic acid and proline which were present in equimolar ratio and which accounted for 30% of the dry weight of TRF" (Schally et al., 1966). In the light of a subsequent change of context this statement was to seem extraordinary (see below). In 1966, Guillemin did not believe Schally's findings. It is also clear however, that Schally did not believe his own findings. Thus, Schally, wrote at the end of his 1966 paper:

> The results are consistent with a hypothesis that TRF is not a simple polypeptide as has been thought previously, but nevertheless our evidence indicated that 3 amino acids are present in this molecule (Schally et al., 1966).

In order to test the hypothesis that TRF was not a peptide, Schally ordered eight synthetic compounds from a chemical company. Each of these compounds contained three amino acids (His, Pro, and Glu) in all possible permutations. Schally tested each compound and when, a few months later, he had failed to find any activity he concluded: "This indicates that the moiety which formed at least 70% of the TRH molecule is essential for biological activity" (Schally et al., 1968).

It is clear that if Schally had not believed Guillemin's hypothesis, he would have found the structure of TRF(H) in 1966. If he had not believed Guillemin's hypothesis, Schally might have concluded that a specific arrangement of the three amino acids was necessary to explain the lack of activity. Similarly, if Guillemin had believed Schally's result, he could also have found the structure in 1966. But when Guillemin referred to Schally's "isolation," it was always with the use of quotation marks. A curious crossing of paths thus occurred. Schally gave up his hypothesis because of Guillemin's suggestion that TRF may not be a simple polypeptide. He was later to regret this: "the field was sorely confused by your strange theory published in . . . that releasing hormones and TRH are not polypeptides" (Schally to Guillemin, 1968).

In 1968, Guillemin "independently" found that three amino acids (His, Pro, and Glu) existed in equimolar ratio and that 80% of the weight was accounted for by amino acids. As a result, Schally resurrected the earlier programme which had almost ceased and relocated his 1966 paper as part of a chronology which supported his claim that he had been right from the beginning. The ambiguity of Schally's retrospective reassessment of his 1966 paper is clearly apparent in the reasons he gave for not immediately following up his 1966 results:

> S: I don't see why we discuss that . . . in 1966 I got the structure . . . everyone agrees on that . . . it's all written. . . .
>
> Qu: But why did you doubt your own results?
>
> S: But I dropped the question. It was of not interest to me. My interest was in reproduction and control of growth hormones . . . . I didn't have a good chemist, I gave it to . . . he was too busy; he had 5000 things to do . . . he never came up with anything, nothing was done for 2 or 3 years.
>
> Qu: But why did you conclude that TRH was not a peptide?
>
> S: Because there was no activity. We believed Guillemin. (Schally stands up, picks up a copy of one of Guillemin's papers and begins to quote from it. . . .)

Qu: Why did you believe in Guillemin's mistake?

S: We never believed in it. . . It's a very difficult thing . . . we
       found impure fractions . . . there was no activity . . . when
       Guillemin came up with his idea of a nonpeptidic moiety we
       followed him. It's something that can always happen (Schally,
       int., 1976).

This example demonstrates that the logic of deduction cannot be
isolated from its sociological grounds. We can say, for example, that
Schally "logically" deduced that TRF was not a polypeptide only if we
simultaneously appreciate that the weight given to Guillemin's theory
was stronger at that time than the weight given to the evidence
produced by Schally. Guillemin was "logical" in concluding that the
enzyme test showed TRF not to be a peptide only in the sense that he
placed more confidence in the enzyme test than in the notion that all
releasing factors are peptides. Following Bloor (1976), we would say
that "logically" possible alternatives were deflected by prevalent
beliefs. For instance, Guillemin eliminated the possibility that his
enzyme test was incomplete. In testing for the activity of different
permutations of synthetic amino acids, Schally eliminated the possibility that changes in the chemical structure of an amino acid might
cause activity. Every modification of context entails the making of
different deductions, each of which will be equally "logical" (see
below). It is thus important to realise that when a deduction is said not
to be logical, or when we say that a logical possibility was deflected by
belief, or that other deductions later became possible, this is done with
the benefit of hindsight, and this hindsight provides another context
within which we pronounce on the logic or illogic of a deduction. The
list of possible alternatives by which we can evaluate the logic of a
deduction is sociologically (rather than logically) determined.

By 1968, a large number of techniques from other disciplines had
been imported into the TRF field, as indicated by the extent of new
citations by TRF papers (see Fig. 3.2). The adoption of the strategy to
"go for the structure at all cost" entailed both the use of techniques
from other disciplines and a resultant modification of the nature of the
research task. Firstly, participants drew upon more established areas
of classical endocrinology in order to obtain reliable bioassays.
Secondly, they borrowed purification techniques from peptide chemistry. This turned out to be relatively easy since Guillemin had already
obtained a 1,000,000-fold purification as early as 1966. Thirdly,
participants amassed a vast quantity of brain extracts (Fig. 3.4).

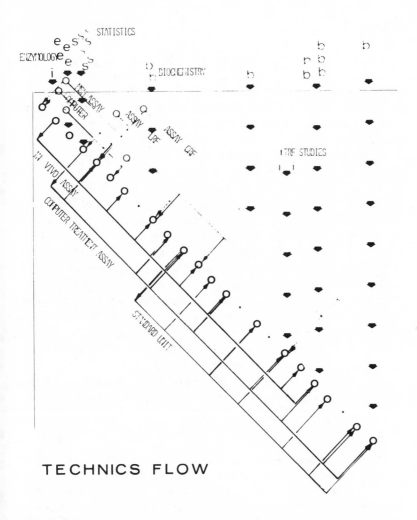

TECHNICS FLOW

Figure 3.4
As in Figure 3.3 this is a simplified representation of the TRF area. This time only Guillemin's papers are shown, and only those operations corresponding to the borrowing of techniques are represented. The continuous arrows indicate the extent to which the group quotes its own articles; the discontinuous arrows indicate from which major fields (and when) the development of the TRF area necessitated import. Once again, the complete mesh of all the operations is a fairly good approximation of the field—as far as papers are concerned. In this case it shows the material layout on which the signs of the existence of TRF could be constructed.

Although exacting, this task required little more than good management and a great deal of patience. This three-fold transformation of the
TRF area greatly raised research standards. Indeed, such was the
degree of required chemical expertise that several competing groups
(groups which in Schally's terms "lacked guts") found themselves
ruined.

At the same time, the adoption of a strategy with an all or nothing
object entailed enormous risks. Even if they obtained highly purified
material, researchers' efforts would count for little if they failed to
determine the structure. The borrowing of techniques of analytical
chemistry entailed the use of expertise and equipment which was
costlier than borrowing techniques of purification chemistry. One
reason for this was the instrumentation of analytical chemistry itself
incorporated many of the advances of physics. In particular, peptide
chemistry had developed powerful tools for determining the structure
of biological substances. However, researchers experienced some
difficulty in relocating themselves within this neighbouring field.
While located within physiology, TRF remained an interesting
substance in that its mode of action could be studied, even though its
structure could not unambiguously be identified. In order to achieve
such identification it was necessary to relocate the substance in the
new context of analytical peptide chemistry. The frustrations met by
researchers in their attempts to achieve this relocation are well
illustrated in the following passage written in 1968:

> Our efforts at characterizing the chemical structure of TRF have led us
> to the conclusion that we are dealing with a rather difficult problem for
> which classical methodology is turning out to be of only limited
> significance. With the preparations of highly purified TRF which we
> have studied so far, the material has appeared to be non-volatile at
> atmospheric pressure which *precludes* the use of gas chromatography,
> or in high vacuum of the order of 10^{-7} torr even at $130°C$, which pre-
> cludes the use of mass spectrometry to study it. The classical derivatives
> which are usually made in these circumstances (methyl, trimethyl sylyl,
> pivalyl) have *not* yet proved to be of any help in studying this problem.
> Nuclear magnetic resonance spectra of highly purified TRF at 60, 100
> or 220 megahertz with time averaging have *not* yielded any meaningful
> information except that we may be dealing with a highly saturated
> alicyclic or heterocyclic structure with peripheral CH_3 groups, without
> completely ruling out a polyamide structure. Infrared and ultraviolet
> spectra have *not* contributed much information either. One of the main

problems here is that of the usually minute quantities of materials available for each one of these methods which are thus stretched to their level
of highest sensitivity with a corresponding loss of specificity of the
information obtained. In view of the extremely high cost of the starting
material and the minute quantities of pure TRF that can be obtained
from this starting material it would appear that the solution to the chemical characterization of the molecule of TRF will require *some of the
most advanced methodology that physical or organic chemistry* is
presently offering or still in the process of developing. . . . More rewarding have been a series of experiments dealing with the physiological
studies with TRF (Guillemin et al., 1968:579).

In other words, it was felt that the initial strategy of going for the
sequence rather than the mode of action might have been a mistake. At
a symposium held in Tokyo and attended by the majority of researchers
in the TRF subfield, there were a number of exchanges between those
convinced of the value of the chemical approach and physiologists
such as Harris, who saw no virtue in committing the whole field to this
task. In 1966 McCann received the Endocrine Society award. This
had the effect of legitimating the classical physiological approach to
the problem, just at the time when both Schally and Guillemin were
bogged down in the most difficult part of their chemical extraction
work.

Many participants were by now aware of the radical differences
implied by the new approach, of the growing competition between the
groups led by Schally and Guillemin and of the extreme difficulty in
effecting a transition between isolation and analytic chemistry.
However, no one was as puzzled by the state of the field as the funding
agencies. For eight years increasing amounts of money had been
invested in the field, but fewer and fewer results had emerged. At the
end of 1968, the situation came to a head when a review committee of
the NIH was set up to assess what was wrong in the field and, in
particular, to assess researchers' chemical expertise and to review the
likelihood of them obtaining the structure (Burgus, 1976; McCann,
1976; Guillemin, 1975; Wade, 1978). Clearly, the principle of laissez
faire was not respected at this time. Researchers in the field were
summoned to Tucson in January 1969 to show where they stood,
under explicit threat of the possible withdrawal of funding and a
subsequent return to the cheaper but more rewarding realms of
classical physiology.

Guillemin, who was just getting new results, did everything he could
to delay this meeting for several months (Guillemin, 1976). In

common with other members of his laboratory, he felt that the public exposure of preliminary results would do more harm than good. By this stage, however, he had begun to collaborate with Burgus, a chemist attracted to the field once the stability of the TRF assay had convinced him that the use of analytic chemistry would not be wasted on an elusive substance, as CRF had turned out to be (Burgus, 1976). Indeed, everything depended on Burgus's chemistry. Guillemin was not a chemist. Schally had stopped working on the programme, and only Burgus could guarantee access to the harder field. It is difficult to assess whether or not the programme would have been discontinued at this point if Burgus had not presented convincing results. The process of accumulating materials and brain extracts had probably passed a point of no return some time in 1968. Nevertheless, access to chemistry might have been prevented by a lack of funds, and a long delay might have occurred if the funding agencies had carried out their threat.

At the Tucson symposium held in January 1969, many participants reported feeling intensely discouraged after the opening sessions. No headway had been made, the chemistry being used appeared somewhat dubious, and there were some open disputes between chemists and endocrinologists. But the situation changed when Burgus began to speak:

> [W]ith the availability of the 1 mg of material which we have just obtained in the last few weeks, we have finally been able to obtain an amino acid analysis:
>   His: 28.5   Glu: 28.1   Pro: 29.2
> . . . these amino acids together make up 80% of the total weight of the preparation (Burgus and Guillemin, 1970a: 233).

This demonstrated that TRF comprised three amino acids in equimolar ratio. In other words, the idea that TRF was not a peptide was probably wrong. As a result, the argument that TRF was not inactivated by enzymes and was thus not a peptide was reversed. A subsequent explanation of the lack of enzyme inactivation portrayed the earlier work as mistaken:

> It is not surprising that the proteolytic enzymes do not act on the molecule considering the three amino acids that are present. We have also considered the possibility of a cyclic or protected peptide being involved which would also explain the resistance to proteases (Burgus and Guillemin, 1970a:236).

However, Burgus stopped short of claiming that TRF was a peptide and nothing else. When asked about this in the following discussion, he stressed the dramatic turn which had taken place when he explained why a follow-up experiment had not yet been carried out: "Our mode of thinking on the polypeptide nature of this material has changed basically in the last 2 or 3 weeks" (Burgus and Guillemin, 1970b: 239). The precise details of the way this change had come about were not immediately apparent. Nevertheless, from the point of view of the sponsors of the symposium, Burgus's results came as a relief. Everyone congratulated the speakers. One of the chemists who had been especially invited to monitor the quality of the chemistry commented:

> I would like to congratulate Drs. Burgus and Guillemin, and also Dr. Schally on two very elegant and exciting chemical papers; I am sure that many of us feel that things are getting rather close now, and the criteria for purity in both cases were extremely impressive (Meites, 1970: 238).

The closeness mentioned here refers to the singular objective, pursued by both Guillemin and Schally, of obtaining the structure of TRF using peptide chemistry. The reference to impressive criteria clearly reflects the increase in standards between one professional boundary and another. Several respondents recalled their feelings of optimism that the field would be saved and that money would not stop.

Bearing in mind our earlier discussion, however, it is not surprising that Schally's reaction was very different. Schally's group contributed little to the published discussion, except to note that, "incidentally we were the first to report (in 1966) that these are the three amino acids in the molecule of TRF" (Meites, 1970: 238). However, his recollections in interview were more vivid:

> But at the Tucson meeting, when I heard the report of Guillemin, my God, I thought we were on the right track all along in 1966. It came as a complete surprise to me . . . we worked like hell . . . then I immediately made a pact with F. (Schally, int., 1976).

In the context established by Burgus's results, Schally's 1966 paper not only became worth believing, it also became the retrospective precursor of the Tucson paper and hence provided his major claim to credit.

Narrowing Down the Possibilities

A bioassay carried out on a partially purified fraction can be thought of as a "soft" technique in the sense that each resulting inscription can be interpreted in tens of different ways. By contrast, an amino acid analysis (AAA) is "hard" in the sense that the number of possible statements which could fit each inscription is very much smaller (Moore et al., 1958). The difference between soft and hard techniques does not depend on any absolute evaluation of the quality of techniques. Hardness simply refers to the fact that a particular material layout permits the advanced elimination of many more alternative explanations (see Ch. 6).

In 1962 Guillemin had decided to go for the TRF structure. By 1968, however, he had not yet obtained the single interpretation that this goal necessitated. TRF had become both an active fraction in bioassays and a sizable (1 mg) sample in an amino acid analyzer. The use of analytic chemistry made it possible to believe both that TRF had existed between 1962 and 1968 and that three amino acids were present in the molecule. But TRF could still have been a variety of other things; it might have been histidine, glutamic acid, and proline in any one of six possible combinations; it might also have been a three, six, or nine amino acid sequence (the same sequence being repeated several times); finally, it could merely have been a component of a larger active molecule, since 20% of the weight was still unaccounted for. In other words, although between 1966 and 1969, Burgus had dramatically narrowed the number of possibilities by importing more and more techniques of analytic chemistry, too many remained. At the same time, it was becoming yet more difficult to eliminate the last few possibilities because researchers were nearing the limits of sensitivity of their instruments.

Each new experiment could redefine the range of possible alternative explanations.[11] For instance, what was known about the weight of TRF was compatible with a tri-, a hexa-, or a nona-peptide. Once the weight was taken as reliable, the alternative explanation that TRF was more than a nona-peptide was eliminated because of its incompatibility with this fact. From another point of view, however, the range of alternative explanations could be increased. Burgus, for instance, did not believe that TRF was merely a peptide, still less that it was a simple tripeptide. Consequently, he delayed his final choice by considering a larger number of possibilities than eventually turned out to be necessary. In a similar way, each new method, each fresh exchange with colleagues and each change in the evaluation of colleagues' beliefs either widened or narrowed the range of possible alternatives. At the Tucson meeting, considerable excitement was engendered by the sudden realisation that after seven years of unrewarding work, the number of possible characterisations of TRF was dramatically reduced. In 1962 TRF might have consisted of any of the combinations of the twenty amino acids then known; by 1966, the range of alternatives had been increased—TRF might also have been some other possible arrangement of a nonpeptidic nature. Suddenly, in 1969, it could only be *one* of twenty or thirty possibilities. In the seventy years of analytical chemistry, the strategy used to attain one such possibility was to talk of the substance in terms of its primary structure (Lehninger, 1975).

The ultimate objective was to attain the particular structure of TRF. It was an ultimate objective because, once attained, a synthetic replica could be produced and compared with the original substance. It was also ultimate in the sense that, given the choice of strategy, nothing further remained to be known. Aristotle defined a substance as something more than its attribute. In chemistry, however, a substance can be so completely reduced to its attribute that an exactly similar substance can be obtained de novo (Bachelard, 1934). This in part explains participants' fascination for the objective. If the exact structure could be obtained, some of the solidity of chemistry and molecular biology could be injected into endocrinology. Or, at least, one unknown ("what exactly is it that we inject?") could be eliminated and the sophistication of all subsequent bioassays could be enhanced.

The requirements for stabilising the structure of TRF were simple: Traces obtained from inscription devices had to be transposed into the language of chemistry. It was known that only three amino acids were present in the substance and that only one arrangement of these acids could trigger activity. The difficulty of identifying the particular arrangement of the amino acids in 1969 is illustrated by Table 3.1. Each of the structures proposed resulted from the application of a new method to the problem and each survived only a few months. Obviously, it is necessary to show precisely how this flux of different names stabilized into one single sequence.

One indirect means of determining the sequence was to synthesize all six possible combinations of the three amino acids known to exist in equimolar ratio in TRF. As we saw above, Schally did this in 1968 but failed to find any activity. Burgus followed the same approach in 1969 and similarly found that none of the synthetic peptides exhibited activity. By 1969, however, the context had changed. Instead of concluding, as Schally had done two years earlier, that TRF was not a peptide, Burgus's negative results were taken as evidence that "something should be done to the N terminal." This involved further chemical manipulation of all six peptides. As a result of one such manipulation, known as "acetylation," it was found that one and only one peptide showed activity: "it seems that the sequence R-Glu-His-Pro is necessary to biological activity, instead of any distribution of the three amino-acids" (Burgus et al., 1969: 2116).

The synthetic replica of TRF was thus known before knowledge of the natural TRF had been constructed. In other words, the use of synthetic chemistry was sufficient to narrow down the possible sequences of TRF from six to one, without having to touch the precious micrograms of the natural extract.

This operation, however, demonstrated only that the synthetic material R-Glu-His-Pro was biologically active, not that natural TRF had the structure R-Glu-His-Pro. To demonstate this further point, inscriptions obtained both from natural and from synthetic material had to be compared. Schally's group tried to do this by comparing the thin layer chromatographs (TLC) of the two substances in twenty different systems. But this was not regarded as an acceptable proof in Guillemin's laboratory. Whether or not the number and quality of inscriptions constituted a proof depended on negotiations between members. It was extremely difficult to decide whether or not two chromatographs (one for synthetic and one for the natural sample) were similar. Having evaluated small differences as meaningful, Burgus wrote: "Given the *difference* of specific activity and of behaviour in several chromatographic systems, it was *obvious* that Pyro-Glu-His-Pro-OH was *not identical* to native TRF" (Burgus et al., 1969b: 226). He went on to propose a further modification which would reduce the remaining small differences and thus allow the specification of one sequence for TRF: "One of the most interesting structures would be Pyro-Glu-His-Pro-amide, because there exist a great number of biologically active polypeptides with a C terminal which is amidated" (Burgus et al., 1969b: 227).

The notion that the peptide could also be amidated led to the fabrication of a compound which would reduce the difference between the two sets of observations on the chromatograph. Indeed, once synthesized, this new compound was found to be similar to natural TRF both in bioassays and in other inscription devices: "[T]he properties of TRF most closely *matched* that of the amide, *failing to separate* from the synthetic compound in four different systems of TLC when run in mixtures" (Burgus et al., 1970).

It would be inadequate for us simply to conclude that TRF is or is not Pyro-Glu-His-Pro-NH$_2$. Difference or identity do not exist per se; rather they depend on the context in which they are used and on negotiations between investigators. It was thus possible either to dismiss a difference as minor noise or to deem it a major discrepancy. Guillemin's group observed "slight differences" between natural and synthetic compounds as revealed by various inscription devices. So seriously did they view these slight differences, however, that in the paper published in July they wrote, "Thus, the structure of TRF is *not* Pyro-Glu-His-Pro-OH, Pyro-Glu-His-Pro-OMe nor Pyro-Glu-His-Pro-NH$_2$" (Burgus et al., 1969b: 228) But for this statement there would have been no subsequent dispute over the allocation of credit and the story would have ended in July 1969.[12]

While Guillemin's group were considering more possibilities than turned out to be necessary, Schally's group published two papers (written by Folkers and submitted on August 8, 1969 and September 22, 1969). In these papers, neither the revelation at the Tucson meeting nor the period between 1966 and 1969 are mentioned. Instead, the 1966 paper was portrayed as the first in which the correct amino acid analysis had been given. The first of Folker's 1969 papers, entitled "Discovery of Modification of the Synthetic Tripeptide Sequence of the TRH Having Activity," refers to Pyro-Glu-His-Pro-NH$_2$ as *one of several* active peptides. Guillemin, however, claimed that this idea had been passed from one group to another during an informal talk at the June (1969) meeting of the Endocrine Society. It is as difficult to establish the truth of this claim as it is of Schally's response (private communication, 1976) that he already knew of this modification but had been "instructed not to tell." The second of Folker's 1969 papers, entitled "Identity of TRH and Pyro-Glu-His-Pro-NH$_2$" (Boler et al., 1969), records Folker's decision to deem *identical* the natural and synthetic substances. To fortify his claim to priority, Folker cited Burgus's paper: "Burgus *et al* (1969b) states that the structure of ovine TRH *is not* that of Pyro-Glu-His-Pro-NH$_2$ and that a secondary or tertiary amide modification is not excluded" (Boler et al., 1969: 707). Curiously, however, Boler et al.

appear to contradict this statement in the next paragraph of the same paper: "If the structure of TRH is not that of Pyro-Glu-His-Pro(NH_2), then certain possibilities are evident" (Boler et al., 1969: 707). In other words, Folkers toyed with alternative structures of TRF even though his paper's title indicated that he had definitely settled on one. This is a good example of what the style of a paper can achieve. Schally's statements allowed Guillemin's group to accuse Schally's group of double talk. As far as Guillemin's group was concerned, Schally did not have greater proof than they that Pyro-Glu-His-Pro-NH_2 was the structure. Rather, they saw Schally's statements both as an expression of confidence in Burgus's argument and as a means of beating the "overcautious" Burgus by two months. As we have shown above, Burgus could not rely on Schally but had to establish fresh sources of information.

Given the organisation of peptide chemistry at this time, Burgus considered that only mass spectrometry could provide a fully satisfactory answer to the problem of evaluating the differences between natural and synthetic TRF. Once a spectrometer had been provided, no one would argue anymore.[13] The strength of the mass spectrometer is given by the physics it embodies. It is not our purpose here to study the social history of mass spectrometry. Suffice to say that for a peptide chemist its use constituted the ultimate argument because, as Burgus (1976) put it: "it eliminates all but a few possibilities." The use of chromatographs alone could enable chemists to continue arguing that the structure of TRF might be different and to propose alternative interpretations. Thus Burgus (1976) made the following comment about Schally's use of thin layer chromatography (TLC): "any *good* chemist will tell you that TLC does *not* make a proof." The only way to avoid further argument and to settle the question was through mass spectrometry. Whereas similarity between the traces of synthetic and natural material could be taken to be coincidental in other systems, mass spectrometers provided information at the level of atomic structure. Although there might be thousands of ways of explaining similar activity in an assay, or in a chromatograph, there were only very few possibilities for explaining similarity in a mass spectrometer. Burgus therefore forecast that whoever obtained the spectra of natural and synthetic TRF would settle the question for ever (see Table 3.1).

Unfortunately, the use of mass spectrometry had thus far been limited because the sample of TRF was not volatile. Without the means of making samples volatile, the final unambiguous determina-

Table 3.1

Before 1962	Is there a TRF?		
After 1962	There is a TRF.		
	What is it?	It is a peptide.	
Around 1966		It might not be a peptide.	
		It is not a peptide.	
January 1969		It is a peptide.	It contains His, Pro and Glu.
April 1969			It is either R-Glu-His-Pro
			or R-Glu-His-Pro-R
			It is not Pyro-Glu-His-Pro-OH
			nor Pyro-Glu-His-Pro-OMe
			nor Pyro-Glu-His-Pro-NH$_2$
November 1969			TRF is Pyro-Glu-His-Pro-NH$_2$

tion of structure could not be made. Consequently, there ensued a period of several months during which investigators tried several ways of inserting the sample into the mass spectrometer in such a way that it became volatile. "This is not a major technological advance, but it is one made for this particular program . . . that is why it took so long, we had to stop and develop this technique" (Burgus, 1976).

Finally, Burgus was able (sometime in September 1969) to introduce the natural sample into the mass spectrometer, and to obtain a spectrum that no one in the field could interpret as being significantly different from that for the synthetic material: "This is the first instance of the structure of a natural product being determined on the basis of its similarity with a synthetic product" (Burgus and Guillemin, 1970).

Here we reach a turning point in the TRF story. Researchers in the TRF field no longer said that natural TRF had a spectrum "similar to" Pyro-Glu-His-Pro-NH$_2$, nor that TRF was "like" the synthetic compound Pyro-Glu-His-Pro-NH$_2$. Instead, a major ontological change took place (see Ch. 4). Participants were now saying that TRF *is* Pyro-Glu-His-Pro-NH$_2$. The predicate became absolute, all modalities were dropped and the chemical name began to be the name of a real structure. Immediately, the status of TRF was transformed into that of a fact, and statement such as "Guillemin and Schally *have established that* TRF is Pyro-Glu-His-Pro-NH$_2$" became commonplace.

TRF Moves into Other Networks

The pure fraction of TRF obtained by the use of highly sophisticated tools of analytic chemistry could be identified simply in terms of a string of eight syllables. This label will remain unambiguous as long as analytical chemistry and the physics of mass spectrometry remain unaltered. The advantage of having situated TRF in the relatively restrictive context of analytic chemistry became obvious as early as November 1969. To find out what TRF was before this date would have entailed a laborious search through a complex mesh of forty-one papers, full of contradictory statements, partial interpretations, and half-baked chemistry. After November 1969, however, eight syllables enabled the rapid spread of news by telephone or by word of mouth and thus raised the possibility of a radical change of network structure. A tiny group of specialists might have concerned themselves with the same problem for years, simply by citing a relatively small number of papers. Now, however, a considerably larger public could use the eight syllable formula as a fresh starting point for their research. The three amino acid formula also had the substantial advantage that it could be used to order as great a quantity of the substance from any chemical company as money was available to pay for it.

The crucial point we have tried repeatedly to stress in this chapter is that once one and only one purified structure had been chosen out of all the equally probable structures, a decisive metamorphosis occurred in the nature of the constructed object. A few weeks after the stabilisation of TRF, nonproblematic samples of the purified material began to circulate within circles of researchers far removed from the original groups led by Guillemin and Schally. These circles comprised groups and laboratories which the impure, problematic fractions (found only to be active in cumbersome and unreliable assays) would never have reached. For these new groups TRF rapidly become taken for granted. Its history began to fade away, and remaining traces and scars of its production become less and less significant for practising scientists. Instead, TRF became just one more of the many tools utilised as part of long research programmes.

The difference between the eight years of effort and the simplicity of the final structure of three amino acids; the disproportion between the tons of hypothalami that were processed and the mere micrograms of substance which were eventually obtained; the fierce competition between the two groups; the drama of the Tucson meeting—all these features enabled TRF to take on a new significance within yet another network—that of the press. TRF became a story and the use of tons of sheep's brain a myth. People who had been totally disinterested by the production of forty-one articles over ten years could now become interested in the final event which they, in turn, helped to highlight and dramatise.[14]

NOTES

1. We use the term in the sense developed by Bloor (1976). Our particular interest is with the aspect of the strong programme which Bloor refers to as "impartiality" (1976: 5). However, our contention is not just that sociological explanation should be impartial with respect to truth or falsity, and that both sides of the dichotomy require explanation. Our argument is that the implicit (or explict) adoption of a truth value alters the form of explanatory account which is produced.

2. Since the award to one of our informants of the Nobel Prize for medicine for this episode, a number of journalistic accounts have appeared. It is of interest to compare the present account with these. See especially Wade (1978) and Donovan et al. (forthcoming).

3. The figures used here are based on three sources: firstly, we used the publication lists of the two main groups engaged in the work; secondly, we recorded all the references in these articles; thirdly, we checked the resulting corpus for completeness against *Index Medicus* and *Permuterm*. All references *to* these papers were obtained either from the SCI or from the other papers in the corpus.

4. The difference between the two expressions also reflects a difference of paradigm. To refer to the substance as a "hormone" means that it is not a new class of substance. Work on "hormones" consequently fits within the classical framework of endocrinology. To call the substance a "factor," on the other hand, permits the integration of the substance in other series of terms (neurotransmitter, for instance) or in a new class by itself (cybernins, for instance) [see, for instance, Guillemin 1976].

5. There are many accounts of this dispute (Wade 1978), some by the participants themselves (Donovan et al., forthcoming). The subject has been treated ad nauseam both in neuroendocrinology and in the press. These accounts concern the kind of obvious social factors which are not of major interest to our argument here: our intention is to analyse the nature of TRF itself. We do not therefore attempt closely to analyse the controversy about chronology. For practical purposes, we follow the California group's accounts more closely.

6. The new restraints that Guillemin imposed on the problem met the approval of the major agencies, especially the American ones. He had already accumulated a large capital of confidence: the monetary capital could be lent him with some certitude of return, even though his demands were very high. For instance in a $100,000 grant application to NIH for buying hypothalami, Guillemin wrote: "A considerable investment has already been made in terms of money, time, and effort in this programme. I consider the present request as a *sine qua non* for its completion" (1965).

7. In the first year, the literature produced by the group included the following: an article describing the "method of calculation and analysis of results of the McKenzie

essay for Thyrotropin," which is a statistical study including details of computer programming; articles describing the "modified McKenzie assay"; "a proposal for a reference standard" to ease comparison with other investigations; and articles on "methods of purification and collection." The set of techniques so gathered constitute the circumstances through which TRF gained some stability of existence (see Fig 3.4 and Ch. 6).

8. The transformation of accent is common in the study of religion but has yet to be carried out in science. Science is discourse, one effect of which is to assert that it speaks the truth. Lyotard (1975) has shown some of these effects: Knorr (submitted) has studied how the work of writing transforms research findings. The "author," the "theory," the "nature," and the "public" are all *effects* of the text. This is especially important in historical *accounts*. See Barthes (1966).

9. We shall return to a discussion of the term belief in Chapter 6. It is not solely a cognitive term. It also refers to the assessment of investments to be made in an area, the type of equipment to be purchased, which kinds of inscription device are most valued, what counts as a proof, and so on. Guillemin defined the area in such a way that when Schally came to set up a laboratory in competition, he had almost exactly to duplicate the organisation of Guillemin's laboratory. The notion of assymmetry of belief needs to be understood with this material background in mind.

10. The nature of citation refers to Chapter 2 (last section) and to the article by Latour (1976). It is clear that this is a crude reflection of the sum total of *operations* made by papers on one another, but even in this rough form they provide a useful indication of the agonistic field.

11. We have to wait until Chapter 6 before considering the notion of "alternative" on firmer ground. It is obvious at this point that the number of alternatives depends on the agonistic field and that the elimination of one or other alternative depends on the relative weight given to any of the inscriptions.

12. Once again it is necessary not to be taken in by the wording of historical discourse. The notion of the *end* of a story (as we showed above) depended on the Guillemin's strategy to obtain the structure; it depended also on the way a statement was qualified by Burgus et al. in their 1969b article and on the numerous accounts that Schally and Guillemin later gave.

13. The mass spectrometer constitutes a *black box*. It is precisely because of this character that it provides most of the hardness of the field (see Ch. 6). The large prototype of the middle thirties has now become a compact and commonplace piece of equipment, which incorporates a computer to carry out most of the initial interpretations. It has been applied to organic chemistry for thirty years, and specifically to peptides as early as 1959. The extension of its use to releasing factors was thus a relatively small step. Given Guillemin's strategy, no other final proof was available. The power of the equipment lies in the fact that the inscription (the spectrum) is obtained by direct contact of the electron flow with the sampled molecules (Beynon, 1960). Although the number of mediations is very great (Bachelard, 1934), each of the indications is black-boxed, and incorporated into a piece of furniture. Consequently, the final result is taken as incontrovertible.

14. See, for example, *Medical World News,* January 16, 1970; *Le Monde,* January 15, 1970. Each of the numerous articles of this period insist on the fierce competition between Schally and Guillemin, and on the clinical importance of their discoveries. The Nobel Prize, awarded in large part because of the TRF story, regenerated a similar rush of stories in the press.

Chapter 4

THE MICROPROCESSING OF FACTS

Our initial visit to the laboratory established the central importance of literary inscription for laboratory activity: the work of the laboratory can be understood in terms of the continual generation of a variety of documents, which are used to effect the transformation of statement types and so enhance or detract from their fact-like status. In the last chapter, our historical examination of the genesis of a single fact demonstrated the influence of laboratory context in delimiting the number of alternative statements which could be made: only by virtue of a crucial shift between one network and another could a particular statement begin circulation as a fact. On the basis of our argument so far, however, it might be argued that we have as yet to penetrate the very essence of scientific activity, that our description of fact construction has left untouched those aspects of scientific activity which have to do with "logic" and "reasoning." In this chapter, therefore, we return to a close examination of the day-to-day activities of the laboratory in order to extend our enquiry into the most intimate aspects of fact construction. We focus on the routine exchanges and gestures which pass between scientists and on the way in which such minutiae are seen to give rise to "logical" arguments, the implementation of "proofs," and the operation of so-called "thought processes."

Our examination of the daily activities of the laboratory entails an interest in the way in which even the smallest gestures constitute the social construction of facts. Put another way, our concern in this chapter is with the microprocesses whereby facts are socially constructed. As we have argued from the beginning, the sense in which we use the term social refers to phenomena other than the obvious influence of ideology (Forman, 1971), scandal (Lecourt, 1976), or macroinstitutional factors (Rose and Rose, 1976). Such factors scarcely exhaust the social character of science. Moreover, there is a danger that whenever these kinds of social factors are not immediately apparent, certain sociologists of science might conclude that the activity they observe does *not* fall within their domain of competence. For example, the history of TRF presented in the last chapter only once revealed the influence of ideology (p. 123); there was evidence only of the indirect influence of career determination (p. 119); and only on three occasions was there any evidence of the influence of institutional factors (e.g., p.139). The sense in which social is used by some sociologists would thus have yielded only a small number of instances of the clear influence of ideology, manifest dishonesty, prejudice, and so on. But it would be incorrect to conclude that the TRF story only exhibits the partial influence of sociological features. Instead, we claim that TRF *is* a thoroughly social construction. By maintaining the sense in which we use social, we hope to be able to pursue the strong programme at a level apparently beyond traditional sociological grasp. In Knorr's terms, we want to demonstrate the idiosyncratic, local, heterogenous, contextual, and multifaceted character of scientific practices (Knorr, in press). We suggest that the apparently logical character of reasoning is only part of a much more complex phenomenon that Augé (1975) calls "practices of interpretation" and which comprises local, tacit negotiations, constantly changing evaluations, and unconscious or institutionalized gestures. Our objective in this chapter is to show that this is the case and that a belief in the logical and straightforward character of science itself arises in the course of these practices of interpretation. In short, we observe how differences between the logic of scientific and non-scientific practices of interpretation are created and sustained within the laboratory.

It is tempting to start from the premise that the nature of scientific activity is essentially different from those practices of interpretation in nonscientific activity. As we shall suggest, however, such temptation arises in part because scientific practices are all too often displayed through the use of terms such as hypothesis, proof, and deduction. The use of such terms renders scientific practice as different, but it is not clear that they are being used other than tautologically. For example, Garfinkel (1967, Ch. 8), in relating Schutz's (1953) description of scientific activity, represents ten criteria of common sense rationality and adds four which can be taken as peculiar to science. One of these four criteria is that scientists look for "compatibility of end-means relationships with principles of formal logic" (p. 267). However, the only difference between this and the corresponding criteria of common sense practice is the appearance in the former of the term formal logic. As a defining feature of science, the notion of formal logic is clearly being used tautologically. Another criteria, "compatibility of the definition of a situation with scientific knowledge" (p. 268), is identical to its daily life counterpart but for the inclusion of the word scientific. Once again a criterial feature of distinctiveness is used tautologically. Although this manoeuvre is relatively common (Althusser, 1974), it is particularly striking when employed by an author such as Schutz, who has the professed aim of describing phenomenologically the actual practice of scientists at work. Observers familiar with notions fed them by epistemologists find it easy to identify instances of honorific discourse in scientists' practical activity. Scientists thus appear to operate scientifically because they are scientists. For our purposes, the problem is that major differences between science and common sense are established as a result of tautological definitions of these differences. Our position is that if such differences exist, their existence must be demonstrated empirically. We therefore try to avoid using epistemological concepts in our portrayal of scientific activity.

Our examination of the microprocesses of laboratory work is based on observations of actual laboratory practice. This material, obtained by virtue of our quasi-anthropological approach, is particularly suited to an analysis of the intimate details of scientific activity. Sharing the daily life of scientists for two years provided possibilities far beyond those afforded by interviews, archival studies or literature searches. We are thus able to draw upon observations of daily encounters, working discussions, gestures, and a variety of unguarded behaviour.[1]

In the first section of this chapter we explore the range of interests and preoccupations apparent in all interactions between members of the laboratory. In particular, we examine the ways in which facts can be created or destroyed during relatively brief conversational exchanges. Secondly, we consider the process whereby the occurrence of this kind of exchanges becomes transformed into accounts about the genesis of "ideas" and "thought processes." Finally, we discuss sources of resistance to the understanding of facts as socially constructed. How can we account sociologically both for the absence of nonindexical statements and for the belief that there is such a thing as a nonindexical statement?[2]

The Construction and Dismantling of Facts in Conversation

One way to examine the microprocesses of fact construction in science is by looking at conversation and discussions between members of the laboratory. For various reasons, it was not possible to tape-record discussions in the laboratory. For a total of twenty-five discussions, however, notes were compiled which include records of the timing, gestures, and intonation. A number of informal discussions, including snatches of conversations at benches, in the lobby and at lunch, were similarly noted. Tape recorders could not be used, so these notes lack the precision necessary for "conversational analysis." Even in their somewhat crude or "tidied" state, however, these discussion notes provide a useful opportunity for a close analysis of the construction of facts.

We began by considering three short excerpts from an informal discussion in order to illustrate some of the ways in which arguments are constantly modified, reinforced, or negated during ordinary interaction in the laboratory. The conversation took place between Wilson, Flower, and Smith in the lobby. Smith was on the point of leaving when Wilson began to talk about an experiment he had done some days previously:

(a) Wilson (to Flower):   You know how difficult this ACTH assay is, for the lower amount. . . . I was thinking, well, for fifteen years I have wasted my money on his assay . . . Dietrich had calculated an ideal curve. Last time he made a mistake, because if you look at the real data, each time ACTH goes down Endorphin goes down, each time ACTH goes up Endorphin goes up. So we are going to calculate the fit between the two curves. Snoopy did it, it's 0.8.

        Flower: Wooh!

> Wilson: And we are going to do it with the means, which
> is perfectly legal. It will be, I am sure, 0.9
> (XII, 85).

Wilson and Flower then began to discuss a paper they were writing for
Science. As Smith again started to leave, however, Wilson turned to
him:

(b) Wilson (to Smith): By the way, I saw on the computer yesterday a
93% (match between) haemoglobin . . . or
yeast?! . . .

     (to Flower): You know what we are talking about? Our
friend Brunick yesterday announced at the
Endocrine Society Meeting that he had an
amino acid analysis for CRF. You know what
happened with his GRF? Smith had a computer programme to look at homologies and
found a 98% homology with haemoglobin, and
I don't know what . . . yeast floating in the
air. . . .

     Flower: That's a case for concern.
Wilson (laughing): Depends on who you are. . . . (XIII, 85).

In the first excerpt, the notion that ACTH and endorphin were the
same was reinforced by the suggestion of a probable improvement in
the fit between two curves. As a result, Smith and Flower were
persuaded that the operation met the desired professional standards.
In the second excerpt, however, a colleague's claim was dismissed by
showing an almost perfect fit between CRF, an important and long
sought-after releasing factor, and a piece of haemoglobin, a relatively
trivial protein. The dismissal effect is heightened by the creation of a
link between his recent claim and the well-known blunder which the
same colleague had committed a few years earlier (cf., Wynne, 1976:
327). Brunick had then claimed to have found a very important
releasing factor, which later turned out to be a piece of haemoglobin.
Brunick's recent claim was severely jeopardised by reference to this
past incident. Flower's subsequent comment ("that's a case for
concern") triggered a response which can be taken as indicating
Wilson's high regard for his own professional standards compared to
those of Brunick.

Smith left when Wilson suggested returning to discussion of the
Science article. Wilson showed Flower a new mapping of the pituitary
vascular system which had been sent to him by a European scientist.
There then ensued a discussion of the map.

(c)   Wilson: Anyway, the question for this paper is what I
                                  said in one of the versions that there was *no
                                  evidence* that there was any psychobehavioural
                                  effect of these peptides injected I.V. . . . Can
                                  we write that down?

                       Flower: That's a *practical* question . . . what do *we
                                  accept* as a negative answer? [Flower men-
                                  tioned a paper which reported the use of an
                                  'enormous' amount of peptides with a positive
                                  result.]

                       Wilson: That *much?*

                       Flower: Yes, so it depends on the peptides . . . but it is
                                  very important to do . . .

                       Wilson: I will give you the peptides, yes we have to do
                                  it . . . but I'd like to read the paper. . . .

                       Flower: You know it's the one where . . . .

                       Wilson: Oh, I have it, OK.

                       Flower: The threshold is 1 ug. . . . OK, if we want to
                                  inject 100 rats (we need at least a few micro-
                                  grams) . . . it's a practical issue (XII, 85).

Unlike previous excerpts, this last sequence shows Wilson asking a
series of questions. Wilson and Flower can be thought to have roughly
the same academic status, even though Flower is about ten years
younger than Wilson. They are both heads of laboratories and
members of the National Academy of Science. However, Flower is an
expert in the psychobehavioral effects of neurotransmitters whereas
Wilson is a newcomer to this field. Wilson therefore needs the benefit
of Flower's expertise in writing the collaborative paper (drafts of
which had already been prepared at the time of the above conversation). More specifically, Wilson wants to know the basis for the claim
that the peptides have no activity when injected intravenously (I.V.),
so that they can counter any possible objections to their argument. At
first sight, a Popperian might be delighted by Flower's response. It is
clear, however, that the question does not simply hinge on the presence
or absence of evidence. Rather Flower's comment shows that it
depends on *what they choose to accept as* negative evidence. For him,
the issue is a practical question. Flower and Wilson follow this
exchange with a discussion of the amount of peptides they require to
investigate the presence of psychobehavioural effects. Wilson had
manufactured these rare and expensive peptides in his own laboratory.
So the question for Flower is what quantity of peptides Wilson is
willing to provide. The discussion between them thus entails a

complex negotiation about what constitutes a legitimate quantity of peptides. Wilson has control over the availability of the substances; Flower has the necessary expertise to determine the amounts of these substances. At the same time, a claim has been made in the literature which could make it necessary to consider using an "enormous" quantity of peptides. In the light of this claim, Wilson's denial that intravenous injection gives a behavioural effect is weakened. On the other hand, Wilson argues that the amount of peptides used in the earlier work is ridiculous because it is far in excess of anything on a physiological scale. Nevertheless, Wilson agreed to give the peptides to Flower and to carry out the investigation with the amount of peptides used by the other researcher. They decided that this was the only way that Wilson's contention could be supported. Significantly, this experiment was planned after Wilson's contention had already been drafted.[3]

Given the context of these discussions, it becomes clear that negotiation between Flower and Wilson does not depend solely on their evaluation of the epistemological basis for their work. In other words, although an idealised view of scientific activity might portray participants assessing the importance of a particular investigation for the extension of knowledge, the above excerpts show that entirely different considerations are involved. When, for example, Flower says, "it is very important to do . . .," it is possible to envisage a range of alternative responses about the relative importance of the uses of peptides. In fact, Wilson's reply ("I will give you the peptides") indicates that Wilson hears Flower's utterance as a request for peptides. Instead of simply asking for them, Flower casts his request in terms of the importance of the investigation. In other words, epistemological or evaluative formulations of scientific activity are being made to do the work of social negotiation.

A single discussion, occupying no more than a few minutes, can thus comprise a series of complex negotiations. The contention that ACTH and endorphin had some common relation was reinforced, Brunick's recent claim was degraded, and work was planned to enhance the resistance to attacks on Wilson's contention about the lack of psychobehavioural effects of certain peptides. These, then, are the results of just some microprocesses of fact construction which take place continually throughout the laboratory. Indeed, the encounter reported above is typical of hundreds of similar exchanges. In the course of these exchanges beliefs are changed, statements are enhanced or discredited, and reputations and alliances between researchers are modified. For our present purposes, the most important characteristic of these kinds of exchange is that they are devoid of statements which are "objective" in the sense that they escape the influence of negotiation between participants. Moreover, there is no indication that such exchanges comprise a kind of reasoning process which is markedly different from those characteristic of exchanges in nonscientific settings. Indeed, for an observer, any presupposed difference between the quality of "scientific" and "commonsense" exchanges soon disappears. If, as this suggests, there are similarities between conversational exchanges in the laboratory and those which take place outside, it is possible that differences between scientific and common sense activity are best characterised by features other than differences in reasoning processes (see Ch. 6).

One evident similarity between the scientific exchanges in the laboratory and those taking place in a nonscientific context is their heterogeneity. Several apparently very different preoccupations feature in exchanges lasting no more than a few seconds. For example, the following exchange took place between two scientists as they were discussing the draft of a paper:

> Smith: I should do the whole sequencing but I don't have enough time.
> Wilson: But these guys from England only put their amino acid analysis in their paper, that's bad manners. . . .
> Smith: And its dangerous because there is definite variance between pig and ovine sequence and you cannot deduce the sequence from the amino acid analysis (IV, 37).

During this exchange Smith and Wilson were sitting at a table, surrounded by drafts, protocol books, and copies of articles. Even though they have already half drafted their paper, the data to support their arguments are not yet available. As Smith comments, the series of investigations necessary to obtain these data would take more time than he could spare. The paper by English researchers which Wilson mentioned (and to which their own paper should necessarily refer) claims that a newly discovered substance A is merely a component part of a known substance B. Since they found that the amino acid analysis of substance A was identical with a portion of the amino acid analysis of substance B (and since they had supplementary reasons to believe that the two substances were related), the English researchers were said to have concluded that the structure of the two substances was the same. Wilson commented that to report the amino acid analysis rather than the sequence was "bad manners." His complaint was that the English researchers had made a claim for the identification of substance A prematurely, where, he (Wilson) was trying to establish the same identification by direct sequencing of substance A. Smith, however, saw the issue as more than just a matter of bad manners. His credibility was at risk because of the danger that a future paper might advance a different structure for substance A, which would make possible accusations that both Smith and the English researchers had prematurely deduced the structure of substance A from its amino acid analysis. This possibility was heightened by participants' knowledge of past attempts to establish structures. By referring to the Dayhoff dictionary of peptides which he kept on his desk, Smith could show that the structure of many substances varied according to the particular species of animal from which peptides are taken. Even so, when he argued that one cannot deduce the structure from amino acid analysis, Smith was not invoking an absolute rule of procedure. In a less risky situation, in a less stringent group, in a case where the dictionary showed no variations, the structure could have been deduced in this way. Since the English researchers had already made this deduction, Wilson and Smith might have been tempted to make the same jump. The decision whether to carry out more experiments or simply to concur that substance A and B were identical thus depended on various evaluations made by Wilson and Smith. For example, whether or not sufficient time was available hinged on Smith's evaluation of the relative importance of other tasks he had to fulfil. The importance of independently deducing the structure depended on Smith's assessment of possible objections in future papers.[4]

These examples of conversations between scientists show that a complex web of evaluations simultaneously enter into any one deduction or decision. In the last example, there were evaluations of the exigencies of professional practice, the constraints of time, the possibility of future controversy, and the urgency of concommitant research interests. The wealth of evaluations makes it impossible to conceive of thought processes or reasoning procedures occurring in isolation from the actual material setting where these conversations took place. Let us now look more closely at the way in which different types of preoccupation enter into exchanges between scientists.

Any utterance can comprise one or more of a number of different preoccupations. Thus, in any given setting, multiple interests may simultaneously enter into any one utterance or utterances may switch rapidly between sets of interests. For example, a series of utterances dealing with what is known about something can suddenly be inter-rupted so that quite different preoccupations come into play. (Who had done that? How good is he?) But these interests are themselves liable to shift abruptly. (Where and what should I publish?) The next utterance might embody yet another preoccupation. (What can we say in this paper?) Moreover, the discussion is always likely to be disrupted by an apparently unconnected issue. (Mike, where did you put the racks?).

A comprehensive typology of interests entering into scientists' discussions would be beyond the scope of the present discussion. Nevertheless, it is possible to discern, albeit in a preliminary way, four main kinds of conversational exchange, each of which correspond to a set of participants' preoccupations.

A first kind of exchange featured reference to "known facts." Discussion of long-established facts was rare and occurred only when this knowledge was thought relevant for contemporary debate. More frequently, discussion about what was known concerned recently established facts. Thus, common kinds of exchanges were: "Eh, has someone already done that?" "Is there a paper on that method?" "When you try this buffer what happens?" When discussion did begin with no references to the past, however, it was not long before parties to the exchange invoked the existence of one particular recently pub-lished paper. The following was part of a lunch time discussion:

Dieter: Is there any structural relation between MSH and Beta LPH?
Rose: It's well known that MSH has parts in common with Beta
          LPH. . . . [Rose went on to explain which amino acids are the
          same. Suddenly, he asked Dieter]: Would you have expected
          finding proteolytic enzymes in the synaptosome?
Dieter: Oh yes.
Rose: Well, has it been known for a long time?
Dieter: Well yes and no . . . there is a paper by Harrison showing that
          they do not obtain (VII, 41).

The exchange began with the kind of statement one would expect to find in a textbook (see Ch. 2). However, the assertion that some-thing is well known was regarded by participants as both insuffi-cient and uninteresting. Rose wanted to know how long this had been well known. Dieter then referred to a paper containing relevant published statements on the matter. Thus, attention was quickly redirected from an item of knowledge itself to an assessment of its nearness to the frontier and its place and time of publication. As a result, the possibility of controversy ("yes and no") was raised. Clearly, these kinds of exchanges serve an information-spreading function which enables group members continuously to draw upon each other's knowledge and expertise to improve their own. These exchanges help to retrieve those practices, papers, and ideas from the past which have become relevant to present concerns.

A second kind of exchange occurred in the course of some practical activity, such as carrying out an assay, when utterances such as the following were common: "How many rats should I use for the control?" "Where did you put the samples?" "Give me the pipette," and "It is now ten minutes since the injection." These are the verbal components of a largely nonverbal body of exchanges during which reference is constantly made to the correct way of doing things. These exchanges take place between technicians, or between researchers and technicians (or between researchers acting as technicians). In their more elaborate forms, these exchanges concern the evaluation of the reliability of a specific method. When, for example, Hills came to the laboratory to talk over a possible collaboration in the isolation of a certain controversial substance, he had to convince researchers of the reliability of the bioassay he had been using. Hills presented details of his method for an hour, during which time he was frequently interrupted by questions:

John: You say methanol . . . is this pure methanol?
Hills: . . . what I think is pure methanol, I don't bother further . . . we use the dish by the seventh day they look like normal cells. They do not differentiate at all and we add a new medium which minimizes growth.
John: We tried that and it works well.
Hills: That's interesting.
Wilson: Is this the ratio you get, John?
Hills: Then when I add — — — plus my substance, there is no response at all.
John: Is this in the *same* dish?
Hills: We flip flop then and after that we will obtain the same response.
John: Hum, that's interesting (VI, 12).

Superficially, this kind of discussion might be thought to be purely technical. As can be shown for the above case, however, there are always a number of undercurrents which constrain both the form and substance of the discussion. For example, John's final expression of interest belied his feeling that Hills's argument was entirely unconvincing. John subsequently reported that he felt unable to probe Hills' argument too severely because he knew that his boss, Wilson, was particularly eager to collaborate with Hills. According to John, his questions were aimed merely at eliminating some fairly obvious objections to Hills's method. Hills's results might have arisen either because the methanol was impure, or because the medium did not minimize growth or because he had used the same dish. John wanted to avoid the possibility of the laboratory's chemists being asked to collaborate with Hills in attempting to isolate a substance which might turn out to be an artefact. In addition, the discussion of Hills's method proceeded with the tacit knowledge of all parties that the substance he had been working on was the focus of a huge grant received by the laboratory several years previously. But despite a grant of several million dollars, their attempts to isolate the substance had so far been unsuccessful. Indeed, according to John, there were already a dozen published claims to have made this isolation, all of which had turned out to be erroneous. The apparently technical discussion of Hills's method thus comprises cautious probing informed by John's evaluation of future collaboration, by the desire to avoid working on an artefactual substance and by the group's current investments.[5]

Occasionally, a third kind of exchange took place. This kind appeared to focus primarily on theoretical matters. By this we mean that there was no obvious reference to the past state of knowledge, to the relative efficacy of different techniques, or to specific scientists and papers. This kind of exchange occurred principally between John and Spencer:

> John: But what you call physiologically meaningful is much larger than what is technically feasible now.
> Spencer: But that's a healthy attitude: it's like defining criteria for neuro-transmitters, it defines future research: by these standards there is no evidence for a physiological role of TRF.
> John: Let's restate the issue . . . originally, I mean phylogenetically, the neurotransmitters are first; the receptors increase all over the place; peptides are just not that evolved: there are less receptors; but I see no difference with neurotransmitters (XIV, 10).

Despite the apparent concern with purely theoretical matters, the above kind of discussion is closely related to other issues. Firstly, the above discussion started because of prior discussion of an abstract which Spencer had to send off that same day. In this abstract Spencer seemed to indicate that TRF was an artefact and of no physiological significance. Secondly, the discussion implicitly related to John's and Spencer's concern about the future of their discipline and the direction that work in their laboratory would take. The shift in the definition of peptidic hormones was important for them: if peptidic hormones were defined as neurotransmitters rather than as classic releasing factors, other methods would have to be used, other collaborations entered into and other research programmes set up. This discussion occurred at a time when TRF had been found to have more and more effects similar to neurotransmitters and was thus in the process of escaping the boundary of the discipline. At the same time, the director of John's and Spencer's laboratory had already shifted his research to psycho-behavioural aspects of substances. If one argues that we are interpreting a theoretical discussion by overemphasising its social background and that this background had been artificially constructed, we can answer that scientists constantly make these kinds of interpretation as part of their evaluation of research programmes.

A fourth kind of conversational exchange featured discussion by participants about other researchers. Sometimes this consisted of reminiscences about who had done what in the past, usually after lunch or in the evening when the pressure of work was relaxed.[6] More common were discussions in which particular individuals were evaluated. This was often the case when reference was made to the argument of a particular paper. Instead of assessing a statement itself, participants tended to talk about its author and to account for the statement either in terms of authors' social strategy or their psychological make-up. For example, Smith and Rickert were discussing an abstract which they had written. In front of them were Rickert's figures, which had been produced by a young postdoctoral researcher working in Rickert's laboratory. Discussion focussed on the abilities of this researcher.

Smith: Are you confident she would be able to do five [more animals]?
Rickert: Her honesty?
Smith: Not her honesty . . . were you confident when she did the others . . .?
Rickert: Oh no, at that level, she is very reliable (IV, 12).

Eventually, Smith and Rickert decided not to proceed with their abstract because they had "more to lose than gain" by publishing results in which they were not completely confident. One of the factors influencing this decision was their evaluation of the personality of the young researcher. It is not clear, however, from Smith's first utterance whether the reliability of the data should be assessed in terms of some personality attribute of the individual concerned. Rickert's response to Smith's first utterance indicates his own confusion.

This kind of reference to the human agency involved in the production of statements was very common. Indeed, it was clear from participants' discussions that *who* had made a claim was as important as the claim itself. (see Ch. 5) In a sense, these discussions constituted a complex sociology and psychology of science engaged in by participants themselves. The following excerpts provide further examples of the ways in which participants' own sociology of science is used as a resource in making decisions and evaluating statements:

I am not particularly anxious to do a grand study with her because she is . . . because of her supercompetitiveness. We will be last on her paper, well twelfth out of fifteen [laughter] (IV, 92).

This occurred during part of a discussion between two participants about whether or not to carry out a particular experiment. The decision to do the work clearly involved an assessment of the kind of strategy likely to be adopted by a collaborator:

They don't know their business. It may be that they see progesterone which has been known for years to be analgesic . . . also, there is a flag in all that. The English have discovered that, they push it. That's normal (VII, 42.).

In a similar way, the above criticism (of a statement made by some English researchers) involves comments about their handling of a discovery.

Although it is possible tentatively to distinguish the above four kinds of conversational exchange, it is also clear that many discussions comprised constant switches from one subject to another. For example, in the course of one discussion (which is too long to reproduce in full), a participant who had just returned from a conference commented that Green had "made an ass of himself." He immediately linked this personal attack to the agnostic statement that "Green is still talking about new, more potent peptides." The speaker then switched to a discussion of techniques in which he related his meeting with Green's chemist:

According to my four hours in the laboratory . . . I was not impressed . . . judging by the published work it is even more embarrassing. . . . Xala [Green's chemist] is Green's Achilles's heel (X, 1).

Thus, in the course of one short discussion references are made to subject matter, to personalities, to claims made at a conference, to techniques used in another laboratory, and to competitors' past claims. After a slight pause, the same speaker added:

So far it is going to change very rapidly, we are the only ones to have antibodies for this substance . . . we seemed to be the only ones doing meaningful work (X, 10).

In this short addition, the speaker links a material element of the laboratory (antibodies) both to the agonistic field and to his own work.

The same excerpt further shows the multitude of interests which enter into discussions, once the two other participants began to talk:

A: We have an interesting thing for you . . . we gave a single dose of B; killed the animals by microwave . . . of course we have some controls without any injection

B: Hum, hum.

A: and we assay them for Beta and Alpha.

B: The whole brain?

A: Yes, and our big surprise was that two and a half hours later

B: [writing carefully] Two and a half hours. . . .

A: it was still 40% the value of Beta . . . the values are here [pointing out a scribbled sheet of paper]. . . .

B: Now this is unbelievable!

A: Of course, the Beta assay is not perfect but we can trust. . . .

B: I think in this case the misreading of Beta cannot be important. . . .

A: No, no, I think

B: [looking at the sheet] Is this point statistically different?

A: Oh yes, I have done it . . . anyway it is different from the control. . . .

B: What is the control?

A: The control is a brain extracted in the same way . . . but we may say something, in the control there is 25 times more Beta than Alpha.

B: That much is already getting interesting.

A: The value is. . . .

B: It's too late to send an abstract to the Federations?! (X, 20).

This exchange took place as participants were looking at a number of data sheets. Expressions such as "this is unbelievable" and "big surprise" stemmed from the expectation that the peptide Beta would degradate quickly and from the contrary indication of the data. B's use of the word "interesting" towards the end of the extract can be understood against a background of controversy over whether either of Beta or Alpha are artefacts. Each of B's questions anticipated a basic objection to the results of the assay. The ability to answer or anticipate these questions depended entirely on the local setting. In other words, it was possible that the assay was unreliable; or that readings resulted from the presence of some other substance. Parties to the exchange thus engaged in manipulating their figures, considering possible objections, assessing their interpretation of statements, and evaluating the reliability of different claims. All the time they were ready to dart to a paper and to use its arguments in an effort not to fall prey to some basic objection to their argument. Their logic was not that of intellectual deduction. Rather, it was the craft practice of a group of discussants attempting to eliminate as many alternatives as they could envisage. By virtue of these microprocesses they attempted to force a statement in one particular direction. In the above case, the notion thought to explain the obtained results (the so-called uptake theory) lasted only three days. Subsequently, the results mentioned by B were explained as having arisen from an artefactual effect.

The comprehensive analysis of all conversations noted in the course of our enquiry would go far beyond the scope of our present argument. It is clear, however, that conversations between practising scientists provide a potentially fruitful source of data which has thus far been largely neglected in studies of scientific practice. Let us therefore summarise some of the opportunities which this material affords. Firstly, conversational material exhibits quite clearly how a myriad of different types of interests and preoccupations are intermeshed in scientists' discussions (Fig. 4.1). Secondly, we have presented evidence to indicate the extreme difficulty of identifying purely descriptive, technical, or theoretical discussions. Scientists constantly switch between interests within the same discussion. Furthermore, their discussions can only be explained in the context of the interests that inform their exchanges. Thirdly, we have suggested that the mysterious thought process employed by scientists in their setting is not strikingly different from those techniques employed to muddle through in daily life encounters. Of course, much more detailed argument is needed to sustain this point satisfactorily. For now, we shall merely suggest that the encounters we have described can be adequately accounted for using the notion of fact construction, and that this makes unnecessary the use of ad hoc epistemological explanations.

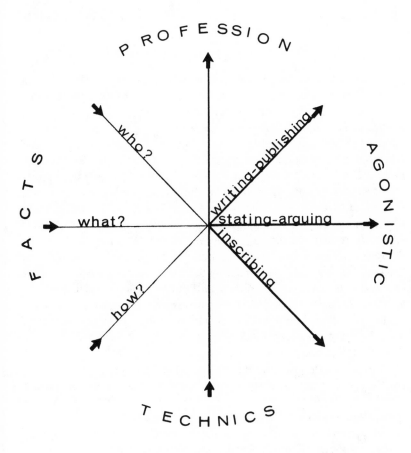

Figure 4.1
This diagram represents the different preoccupations of the conversations we observed in the laboratory. Any utterance can be situated at the middle of the intersecting lines and is susceptible to switch abruptly to any set of preoccupations indicated here. The main sets are the already constructed facts (stage 4 or 5), the individual makers of these facts, the set of assertions in the process of fabrication (stages 1 to 3), and, lastly, the body of practices and inscription devices allowing operations to be performed. Any utterance is thus the integration of these many evaluations. It is in this sense we can say that a scientific assertion is socially constructed.

The Sociological Analysis of "Thought Processes"

Unlike many of the written records of the laboratory, informal discussions provide material which has neither been corrected nor formalised. It is perhaps not surprising that such material provides a wealth of evidence of the intrusion of social factors in the day-to-day exchanges between scientists. But is it possible now to extend the analysis to the realm of thought itself? We have tried to persuade the reader to follow our move from macrosociological concerns to a study of the laboratory and from there to the microsociological study of a single fact. In the previous section we examined how fact construction is affected by conversational exchanges. But the analysis of thought is surely beyond the scope of sociological investigation! It could be argued, for example, that the solitude of the individual scientist in thought excludes the sociologist by definition. Social factors are self-evidently absent from the activity of thought. In addition, it would be argued, the sociological observer is prevented from demonstrating the social character of thought because he is unable to present any written record of thought processes.[7]

Although it might seem wiser to stop sociological enquiry at the level of mute individual thought and to leave the ground to psychology (Mitroff, 1974), psychoanalysis, or to scientists' reminiscences (Lacan, 1966), this would be inconsistent with our argument so far. If we cannot account in sociological terms for scientists' thoughts the ad hoc concepts of which we have tried to rid ourselves will merely take refuge in the "intimate thought process." As a result, science will once again appear extraordinary. Our position is not unlike the opponents of vitalism in nineteenth-century biology. No matter what progress was made by biologists to explain life in purely mechanistic and materialist terms, some aspects always remained unexplained. There were always some corners in which notions of "soul" or the "pure vital force" could find refuge. Similarly, the notion that there is something special about science, something peculiar or mysterious which materialist and constructivist explanations can never grasp, is pushed further and further. But this notion will remain as long as the idea lingers that there is some peculiar thinking process in the scientist's mind. It is to complete our argument and to hamstring efforts to rescue the exotic view of science that we need tentatively to embark upon this new level of microprocessing.

We have already said that a major obstacle to the study of thought processes is the absence of written records. The situation is fortunately more complicated than that, as can be seen from the following account, provided by a member of a nearby laboratory:

Slovik proposed an assay but his assay did not work everywhere; people could not repeat it; some could, some could not. Then one day Slovik *got the idea* that it could be related to the selenium content in the water: they checked to see where the assay worked; and indeed, Slovik's idea was right, it worked wherever the selenium content of water was high (XII, 2).

Clearly, this account is amenable to the kind of treatment found in biblical exegesis (Bultmann, 1921). It is an anecdote of the type "one day so and so had an idea," which, as historians of science well know, is common among the recollections of scientists. The observation that it is an anecdote has an important consequence. Instead of marvelling at how Slovik could have such a good idea and how he could be so damned right, it is possible to formulate an alternative account using sociological arguments based on interview material. This kind of account takes the following form: Firstly, because of an institutional (University of California) requirement that graduate students were obliged to obtain credits in a field totally unrelated to their own, one of Slovik's young students, Sara, had taken selenium studies. She had opted for this because it had a vague relation to her major option. Secondly, there had been a strong group tradition that informal seminars be held where graduate students were asked to talk about unrelated areas in which they had obtained extra credits. Thirdly, at one meeting Sara had presented a paper on selenium dealing both with tissues of interest to her fellow immunologists and with more unrelated questions, such as the influence of selenium water content on cancer. Slovik was at this meeting. A few years earlier he had proposed a cell culture assay, which nobody could reproduce at first, but which was subsequently found to work in some places but not others. The dependence of the efficacy of the assay on geographical location was baffling, mainly because of the prevalent working assumption that scientific principles held universally true. Even Slovik's technician found himself unable to make the assay work outside of his own laboratory. It was not until all the necessary materials and equipment were transported from Slovik's laboratory that the assay was found to work. But even this successful attempt to reproduce identical conditions outside Slovik's laboratory did not reveal that the water was the critical factor. Previously, attempts to repeat Slovik's assays had failed, apparently because of the nature of the cells used by other investigators.

Sara mentioned at the end of her presentation that someone on campus had recently suggested that a trace amount of selenium in water can cause some forms of cancer. The suggestion was that there was a coincidence between the geographical distribution of selenium content in water throughout the U.S. and the occurrence of certain types of cancer. Sara said that no one had taken this suggestion seriously. But Slovik took up the notion that the distribution of selenium content in water could explain the selective occurrence of a particular phenomenon at certain locations.[8] His assays only worked "in some places." It was therefore possible that high selenium content corresponded to locations where the assay would not work. Slovik made a hurried telephone call to one of his colleagues who had been unsuccessful in making the assay work: "Listen, I've got an idea. Sara suggested that it might be the selenium in the water. Can you check that?"

Although this second account is as much a constructed tale as the first, there are some notable differences. The main character of the first is Slovik; the second features a graduate student, Slovik, and the perpetrator of the suggested link between selenium content and cancer. The first account focusses on sudden realisation; the second portrays a multiple progression of accidentally related events. The first highlight's an individual's *idea,* whereas the second mentions institutional requirements, group traditions, seminar meetings, suggestions, discussions and so on. More significantly, the first account is included as part of the second.

Slovik told his colleagues that he had got an idea. Clearly, the attribution of credit for the idea depends to a large extent on which particular version is taken as authoritative. Can the idea be truly said to have first occurred to Slovik rather than to Sara? We shall return to a discussion of actors' appropriation of ideas in the next chapter. For present purposes, it is important to note that having an idea (as in the first account) represents a summary of a complicated material situation. Once the connection between selenium content and the assay was made, all the attendant social circumstances disappeared. By transforming the second account into the first, the teller transforms a localised, heteregoneous, and material set of circumstances (in which social factors are clearly visible) into the sudden occurrence of a personal and abstract idea which bears no trace of its social construction.[9]

This example suggests that there may not be any thought process which should be studied by sociologists or psychologists. By this we mean to suggest that individual's ideas and thought processes result from a particular form of presentation and simplication of a whole set of material and collective circumstances. If the observer takes such anecdotes at face value, it will be hard to demonstrate the social character of fact construction. If, however, he treats them as tales which obey certain laws of their "genre," it is possible both to extend the analysis of fact construction and to understand how such stories about ideas and thought are generated.[10]

The above example encourages us to try sociologically to understand what is all too frequently transformed into stories about minds having ideas. A useful maxim is Heidegger's observation that "Gedanke ist Handwerk": thinking is craftwork. An unusually explicit example of the importance of craftwork is related in Watson's (1968) account of the famous Donohue episode. Watson's portrayal of his "pretty model," in which bases are paired along a like-with-like structure, does not situate himself in a realm of thought, but inside a real Cambridge office manipulating physically real cardboard models of the bases. He does not report having had ideas, but instead emphasises that he shared an office with Jerry Donohue. When Donohue objected to Watson's choice of the enol form for picturing the bases, Watson pointed to actual textbooks of chemistry.

> My immediate retort that several other texts also pictured guanine and thymine in the enol form cut no ice with Jerry. Happily, he let out that for years chemists had been arbitrarily favoring particular tautomeric forms over their alternatives on only the flimsiest of grounds (Watson, 1968: 120).

Watson chose to believe Donohue rather than the general opinion expressed in the textbooks for a variety of reasons, not least of which was his evaluation of Donohue's career up to that date.[11] As we shall see in Chapter 5, individuals' careers constitute an important resource for evaluation of their claims. On the basis of his evaluation, Watson cut out new cardboard models of bases and, after moving them about his desk for a while, he saw the symmetry of the cardboard models of the pairs thymine and guanine and adenine and cytosine. If Watson had not written his book, no doubt the complexity of this practice would have been transformed, either into an anecdote that "one day

Watson got the idea of trying the keto form" or into a titanic epistemological battle between rival theories.

A major difficulty for the observer is that he usually arrives on the scene too late: he can only record the retrospective anecdotes of how this or that scientist had an idea. This difficulty can be partially overcome by in situ observation both of the construction of a new statement *and* of the subsequent emergence of anecdotes about its formation. Let us give an example.

In the laboratory, Spencer had been working on neurotensin, substance P, and on analogs of these two peptides. He tried these peptides in several behavioural assays but did not seem very happy with the results. One outcome of this programme, however, was that one analog of substance P, bombesin, seemed closely to match the effects of neurotensin. This was despite the fact that bombesin was totally unrelated to the structure of neurotensin. Some time later, considerable excitement accompanied Spencer's production of a diagram purporting to show the substantial effect of bombesin on the temperature of rats exposed to cold. The unexpected size of this effect attracted much comment in the laboratory. Although bombesin was active in other assays in quantities of a few micrograms, no more than a nanogram was needed to decrease temperature. Members of the laboratory heralded this as a new finding. When asked why he had tried bombesin in an assay which had never previously been used in the laboratory, Spencer replied:

I have been sitting for a long time waiting for someone with a good CNS assay. . . . I tried a lot of things . . . you remember, I tried temperature, tail vibration. I was never satisfied. . . . But temperature that's important. . . . It's easily measureable and directly related to CNS effect . . . Then this paper came by Bis. . . . I really wanted a CNS assay (IX, 68).

The paper written by Bis described the effect of neurotensin on the temperature of rats exposed to cold. On the basis of earlier trials, Spencer knew that bombesin was functionally (but not structurally) related to neurotensin. Consequently, it occurred to Spencer that it might be worth trying out the possibility of a similar effect of bombesin on temperature. Thus, his existing concern with bombesin and his perception of an analogy between the effects of neurotensin and bombesin together prompted him to try out a new effect.[12] As it happened, bombesin was shown to be 100,000 times more active than neurotensin.

In the article subsequently sent to *Science,* the link between bombesin and neutotensin was no longer analogical. Instead, it had apparently been deduced from the importance of bombesin on the central nervous system. But, as we have seen above, this importance was the consequence of the experiment rather than its prior justification. When asked two months later how the link between bombesin and body temperature had been made, Spencer explained that it was a "logical idea . . . it was straightforward, knowing the importance of thermoregulation for frogs" [from which bombesin was originally isolated].

The significance of this example stems not so much from the fact that Spencer modified his account of the discovery over time (Woolgar, 1976; Knorr, 1978), but from the nature of this modification. Initially, the link between bombesin and thermoregulation was weak. The local circumstances of the laboratory made it only a small step between one entity and another. After a while, however, the link became transformed into a strong logical connection. At the same time, the step taken by Spencer appears to have become very large.

The pervasive influence of analogical reasoning will be evident to many observers of scientific activity. Indeed, there exists an extensive literature on the nature of analogy in science (for example, Hesse, 1966; Black, 1961; Mulkay, 1974; Edge, 1966; Leatherdale, 1974). These authors have discussed the kinds of hybridisation process through which new statements are formed and have thereby helped expose the meticulous sorting of weak connections between existing ideas which constitutes the otherwise mysterious act of creation. It has been pointed out that logical connections of the form "A is B" are only one part of a family of analogical connections, such as "A is like B," "A reminds me of B," and "A might be B." Such analogical links have proved particularly fruitful in science even though they are logically imprecise. For example, the syllogism corresponding to the situation described above would take the following form:

Bombesin sometimes acts like neurotensin.
Neurotensin decreases temperature.
Therefore bombesin decreases temperature.

Clearly, this is logically incorrect. Nevertheless, it was sufficient to prompt an investigation which yielded results subsequently acclaimed as an outstanding contribution.[13] Once the new statement has been accepted, the initial premises are modified (through representation in a written or other retrospective account) to make the syllogism formally correct (Bloor, 1976).

Our point is that the kind of work done by scientists and frequently depicted as analogical reasoning is not reasoning. Spencer wanted to carry out a successful assay, he had bombesin in the laboratory and he wanted to make something out of it. He had accumulated data on the similarity of bombesin and neurotensin, he read Bis's paper, and he adopted the assay described by Bis. By reconstructing the material setting, circumstances, and chance encounters, it becomes clear that the decision to try out the effects of bombesin on temperature was a very small step, and far from the audacious logical leap which it was later depicted to be. Precisely because the local circumstances change very quickly, all reference to them disappears once the step has been made. Both participant and observer are soon left with a version of the event which has been eroded of all contingent circumstances. Retrospectively, the two entities (practices or statements) appear unrelated. Consequently, any link between them will appear "outstanding."

We have argued that accounts of the emergence of a new finding (or statement of fact) entail a two-fold process of transformation. On the one hand, the analogical path is often replaced by a logical connection. On the other hand, the complex set of local circumstances which temporarily makes possible a weak link gives way to flashes of intuition. The notion of someone having had an idea provides a highly condensed summary of a complex series of processes. It also forms the basis for an account which begins to come to terms with the essential contradiction between the use by scientists of procedures which are logical (but sterile) and yet fruitful (but logically incorrect). Our argument is not simply that thought processes are readily amenable to sociological study; rather, an important focus of study should be the aspects of scientists' accounting practices through which thought processes are created and sustained.

Facts and Artefacts

The paradox associated with the term fact was spelled out in Chapter 2: fact can have two contradictory meanings. On the one hand, our quasi-anthropological perspective stresses its etymological significance: a fact is derived from the root *facere, factum* (to make or to do). On the other hand, fact is taken to refer to some objectively independent entity which, by reason of its "out there-ness" cannot be modified at will and is not susceptible to change under any circumstances. The tension between the existence of knowledge as pregiven and its creation by actors has long been a theme which has preoccupied philosophers (Bachelard, 1953) and sociologists of knowledge. Some sociologists have attempted a synthesis of the two perspectives (for example, Berger and Luckman, 1971), but usually with somewhat unsatisfactory results. More recently, sociologists of science have convincingly argued the case for the social fabrication of science (for example, Bloor, 1976; Collins, 1975; Knorr, 1978). But despite these arguments, facts refuse to become sociologised. They seem able to return to their state of being "out there" and thus to pass beyond the grasp of sociological analysis. In a similar way, our demonstration of the microprocessing of facts is likely to be a source of only temporary persuasion that facts are constructed. Readers, especially practising scientists, are unlikely to adopt this perspective for very long before returning to the notion that facts exist, and that it is their existence that requires skillfull revelation.[14] In the last part of this chapter, therefore, we discuss the source of this resistance to sociological explanation. It is little use arguing the feasibility of the strong programme of the sociology of knowledge if we cannot understand why it seems systematically absurd to make such an argument. As Kant (1950) advised, it is not enough merely to show that something is an illusion. We also need to understand why the illusion is necessary.

In the case of TRF, we showed when and where the metamorphosis between statement and fact took place. By the end of 1969, when Guillemin and Schally formulated the statement that TRF is Pyro-Glu-His-Pro-NH$_2$, no one was able to raise any further objections to this claim. Laboratories with no interest in the nine-year saga of the emergence of TRF proceeded from this statement merely by citing papers published at the end of 1969. For them the statement was sufficient basis on which to place an order for the synthetic material which promised to decrease the noise of the assays in which they were engaged. From the point of view of the borrowers, the traces of production of the established fact were uninteresting and irrelevant. Five years later, even the names of the "discoverers" of TRF were of no consequence (cf., Fig. 3.2).

We have been careful to point out that our determination of the point of stabilisation, when a statement rids itself of all determinants of place and time and of all reference to its producers and the production process, did not depend on our assumption that the "real TRF" was merely waiting to be discovered and that it finally became visible in 1969. TRF might yet turn out to be an artefact. For example, no arguments have yet been advanced which are accepted as proof that TRF is present in the body as Pyro-Glu-His-Pro in "physiologically significant" amounts. Although it is accepted that synthetic Pyro-Glu-His-Pro is active in assays, it has not yet been possible to measure it in the body. The negative findings of attempts to establish the physiological significance of TRF have thus far been attributed to the insensitivity of the assays being used rather than to the possibility that TRF is an artefact. But some further slight change in context may yet favour the selection of an alternative interpretation and the realisation of this latter possibility. The point at which stabilisation occurs depends on prevailing conditions within a particular context. It is characteristic of the process of fact construction that stabilisation entails the escape of a statement from all reference to the process of construction.

Facts and artefacts do not correspond respectively to true and false statements. Rather, statements lie along a continuum according to the extent to which they refer to the conditions of their construction. Up to a certain point on this continuum, the inclusion of reference to the conditions of construction is necessary for purposes of persuasion. Beyond this point, the conditions of construction are either irrelevant or their inclusion can be seen as an attempt to undermine the established fact-like status of the statement. Our argument is not that facts are not real, nor that they are merely artificial. *Our argument is not just that facts are socially constructed. We also wish to show that the process of construction involves the use of certain devices whereby all traces of production are made extremely difficult to detect.* Let us look more closely at what takes place at the point of stabilisation.

From their initial inception members of the laboratory are unable to determine whether statements are true or false, objective or subjective, highly likely or quite probable. While the agonistic process is raging, modalities are constantly added, dropped, inverted, or modified. Once the statement begins to stabilise, however, an important change takes place. *The statement becomes a split entity.* On the one hand, it is a set of words which represents a statement about an object. On the other hand, it corresponds to an object in itself which takes on a life of its own. It is as if the original statement had projected a virtual image of itself which exists outside the statement (Latour, 1978). Previously, scientists were dealing with statements. At the point of stabilisation, however, there appears to be both objects *and* statements about these objects. Before long, more and more reality is attributed to the object and less and less to the statement *about* the object. Consequently, an inversion take place: the object becomes the reason why the statement was formulated in the first place. At the onset of stabilisation, the object was the virtual image of the statement; subsequently, the statement becomes the mirror image of the reality "out there." Thus the justification for the statement TRF is Pyro-Glu-His-Pro-NH_2 is simply that "TRF *really is* Pyro-Glu-His-Pro-NH_2." At the same time, the past becomes inverted. TRF has been there all along, just waiting to be revealed for all to see. The history of its construction is also transformed from this new vantage point: the process of construction is turned into the pursuit of a single path which led inevitably to the "actual" structure. Only through the skills and efforts of "great" scientists could the setbacks of red herrings and blind alleys be overcome and the real structure be revealed for what it was.

Once splitting and inversion have occurred, even the most cynical observers and committed relativists will have difficulty in resisting the impression that the "real" TRF has been found, and that the statement mirrors reality. The further temptation for the observer, once faced with one set of statements and one reality to which these statements correspond, is to marvel at the perfect match between the scientist's statement and the external reality.[15] Since wonder is the mother of philosophy, it is even possible that the observer will begin to invent all kinds of fantastic systems to account for this miraculous *adequatio rei et intellectus.* To counter this possibility, we offer our observations of the way this kind of illusion is constructed within the laboratory. It is small wonder that the statements appear to match external entities so exactly: they are the same thing.

Our contention is that the strength of correspondence between objects and statements about these objects *stems from the splitting and inversion of a statement within the laboratory context.* This contention can be supported in three ways. Firstly, there are severe difficulties in adequately describing the nature of the "out there-ness" in which objects are said to reside because descriptions of scientific reality frequently comprise a reformulation or restatement of the statement which purports to "be about" this reality. For example, it is said that TRF is Pyro-Glu-His-Pro-NH_2. But the further description of the nature of the TRF "out there" hinges on the repetition of this statement and so involves tautology. Lest the reader thinks this is an unwarranted caricature of the realist position, it is worth quoting from an argument for a "realist theory of science." In essence, the position advocated here is that no theory of science is possible without what are referred to as "intransitive objects of scientific knowledge."

> We can easily imagine a world similar to ours, containing the same intransitive objects of scientific knowledge, but without any science to produce knowledge of them. . . . In such a world, which has occurred and may come again, reality would be unspoken for and yet things would not cease to act and interact in all kinds of ways. In such a world . . . the tides would still turn and metals conduct electricity in the way that they do, without a Newton or a Drude to produce our knowledge of them. The Widemann-Franz law would continue to hold although there would be no-one to formulate, experimentally establish or deduce it. Two atoms of hydrogen would continue to combine with one atom of oxygen and in favourable circumstances osmosis would continue to occur (Bhaskar, 1975: 10).

The author adds that these intransitive objects are "quite independent of us" (p. 21). He then continues with a striking confession: "They are not unknowable, because, as a matter of fact, quite a bit is known about them" (p. 22). Quite a bit indeed! The marvel of the author for the independence of reality belies its initial construction. Moreover, the ontological status accorded these independent objects is enhanced by the vague terms in which they are described. For example, the statement that "metals conduct electricity in the way they do" implies a complexity beyond the scope of present discussion and, by implication, available only to concerted efforts toward the pursuit and revelation of the reality which gives rise to the description provided here.[16] The author can only recall the reality of the Widemann-Franz laws through the use of eponymy. In addition, he wisely confines his discussion to physics, and to pre-Newtonian physics at that. Perhaps the "independence"of "intransitive objects of scientific knowledge" would seem less unproblematic in relation to more recently constructed phenomena, such as chromosomes or non-Newtonian physics. The realist position, exemplified by the above, centres on a tautological belief whereby the nature of independent objects can only be described in the terms which constitute them. Our preference is for the observation of the processes of splitting and inversion of statements which make these kinds of beliefs possible.

Scientists themselves constantly raise questions as to whether a particular statement "actually" relates to something "out there," or whether it is a mere figment of the imagination, or an artefact of the procedures employed. It is therefore unrealistic to portray scientists busily occupying themselves with scientific activity while leaving debates between realism and relativism to the philosophers. Depending on the argument, the laboratory, the time of year, and the currency of controversy, investigators will variously take the stand of realist, relativist, idealist, transcendental relativist, sceptic, and so on. In other words, the debate about the paradox of the fact is not the exclusive privilege of the sociologist or philosopher. It follows that attempts to resolve the essential differences between these positions is merely to engage in the same kind of debates as the subjects of study, rather than to understand how debates get resolved and positions taken as practical and temporary achievements. As Marx (1970) put it:

> [T]he question of knowing if human thought is able to reach an objective truth is not a theoretical but a practical question. It is by practice that man ought to prove the truth, that is, the reality and the power of the something *beyond* his thought.

An important task for the sociologist is to show that the construction of reality should not be itself reified. This can be shown by considering all stages of the process of reality construction and by resisting the temptation to provide a general explanation for the phenomenon.

Perhaps the most forceful argument for the occurrence of splitting and inversion is the existence of artefacts. A modification in the local context of the laboratory may result in the use of a modality whereby an accepted statement becomes qualified or doubted. This yields perhaps the most fascinating observation to be made in the laboratory—the *deconstruction* of reality. The reality "out there" once again melts back into a statement, the conditions of production of which are again made explicit. We have already given a number of examples of this deconstruction process (see for example p. 129ff). The existence of a moiety for TRF was taken as fact for a few years and was almost regarded as reality before it faded away and was found to be an artefact of the purification process. Sometimes the status of statements changed from day to day, even from one hour to the next. The factual status of one substance, for instance, varied dramatically over a period of a few days.[17] On Tuesday, a peak was thought to be the sign of a real substance. But on Wednesday the peak was regarded as resulting from an unreliable physiograph. On Thursday, the use of another pool of extracts gave rise to another peak which was taken to be "the same." At this point, the existence of a new *object* was slowly solidifying, only to be dissolved again the following day. At the frontier of science, statements are constantly manifesting a double potential: they are either accounted for in terms of local causes (subjectivity or artefact) or are referred to as a thing "out there" (objectivity and fact).

While one set of agonistic forces pushes statement towards fact-like status, another set pushes it toward artefact-like status. This is exemplified by the kind of exchange quoted at the beginning of the chapter. The local status of a statement at any time depends on the resultant of these forces (Fig.4.2). The construction and dismantling of the same statement can be monitored by direct observation, so that what was a "thing out there" can be seen to fold back into a statement which is referred to as a "mere string of words," a "fiction," or an "artefact" (Latour, 1978). The importance of observing the transformation of a statement between fact-like and artefact-like status is obvious: if the "truth effect" of science can be shown both to fold and unfold, it becomes much more difficult to argue that the difference between a fact and and artefact is that the former is based on reality while the latter merely arises from local circumstances and psychological conditions. The distinction between reality and local circumstances *exists only after* the statement has stabilised as a fact.

To summarize the argument in another way, "reality" cannot be used to explain why a statement becomes a fact, since it is only after it has become a fact that the effect of reality is obtained. This is the case whether the reality effect is cast in terms of "objectivity" or "out thereness." It is *because* the controversy settles, that a statement splits into an entity and a statement about an entity; such a split never precedes the resolution of controversy. Of course, this will appear trivial to a scientist working on a controversial statement. After all, he does not wait in hope that TRF will pop up at a meeting and finally settle the controversy as to which amino acids it comprises. In this work, therefore, we use the argument as a methodological precaution. Like scientists themselves we do not use the notion of reality to account for the stabilisation of a statement (see Ch. 3), because this reality is formed as a consequence of this stabilisation.[18]

We do not wish to say that facts do not exist nor that there is no such thing as reality. In this simple sense our position is not relativist. Our

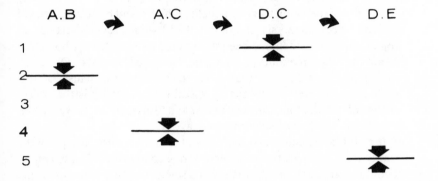

Figure 4.2
If it is assumed that the name of the scientific game is to push a statement (A.B) as far as possible toward fact-like status (stages 4 and 5), then, depending on the resistance encountered (in the form of efforts to transform the statement into an artefact), a scientist has to modify his statement until he can push it to stage 5. The hypothetical example here illustrates the double movement of push and jump. If resistance is too severe, a new statement is forged by an analogical jump and pushed again into the agonistic field. The resultant of this double movement is a drift which follows a pattern peculiar to each statement.

point is that "out-there-ness" is the *consequence* of scientific work rather than its *cause*. We therefore wish to stress the importance of *timing*. By considering TRF in January 1968, it would be easy to show that TRF is a contingent social construction, and moreover, that scientists themselves are relativists in that they are very aware of the possibility of their constructing a reality which could be an artefact. On the other hand, analysis in January 1970 would reveal TRF as an object of nature which had been discovered by scientists, who, in the meantime, had metamorphosed into hardened realists. Once the controversy has settled, reality is taken to be the cause of this settlement; but while controversy is still raging, reality is the consequence of debate, following each twist and turn in the controversy as if it were the shadow of scientific endeavour.[19]

It could be objected that there are other grounds for accepting the reality of a fact apart from the cessation of controversy. For example, it could be argued that the efficacy of a scientific statement outside the laboratory is sufficient basis for accepting its correspondence with reality.[20] A fact is a fact, one could say, because it works when you apply it outside science. This objection can be answered in the same way as the objection about the equivalence of a statement with the thing out there: observation of laboratory activity shows that the "outside" character of a fact is itself the *consequence* of the laboratory work. In no instance did we observe the independent verification of a statement produced in the laboratory. Instead, we observed the *extension* of some laboratory practices to other arenas of social reality, such as hospitals and industry.

This observation would be of little weight if the laboratory was concerned exclusively with so-called basic science. However, our laboratory had many connections with clinicians and with industry through patents.[21] Let us consider one particular statement: "somatostatin blocks the release of growth hormones as measured by radioimmunoassay." If we ask whether this statement works outside science, the answer is that the statement holds in every place where the radioimmunoassay has been reliably[22] set up. This does not imply that the statement holds true *everywhere*, even where the radioimmunoassay has *not* been set up. If one takes a blood sample of a hospital patient in order to determine whether or not somatostatin lowers the level of the patient's growth hormone, there is no way of answering this question without access to a radioimmunoassay for somatostatin. One can *believe* that somatostatin has this effect and even claim by induction that the statement holds true absolutely, but this amounts to a belief and a claim, rather than to a proof.[23] Proof of the statement necessitates the extension of the network in which the radioimmuno-assay is valid, to make part of the hospital ward into a laboratory annex in order to set up the same assay. It is impossible to prove that a given statement is verified outside the laboratory since the very existence of the statement depends on the context of the laboratory. We are not arguing that somatostatin does not exist, nor that it does not work, but that it cannot jump out of the very network of social practice which makes possible its existence.

There is nothing especially mysterious about the paradoxical nature of facts. Facts are constructed in such a way that, once the controversy settles, they are taken for granted. The origin of the paradox is in the lack of observation of scientific practices; when an observer considers that the structure of TRF is Pyro-Glu-His-Pro-NH_2 and then realises that the 'real' TRF is also Pyro-Glu-His-Pro-NH_2, he marvels at this magnificent example of correspondence between man's mind and nature. But closer inspection of the processes of production reveals this correspondence to be much more earthy and less mysterious: the thing and the statement correspond for the simple reason that they come from the same source. Their separation is only the *final stage in the process of their construction.* Similarly, many scientists and non-scientists alike, marvel at the efficacy of a scientific fact outside science. How extraordinary that a peptidic structure discovered in California works in the smallest hospital in Saudia Arabia! For one thing, it only works in well-equipped clinical laboratories. Considering that the same set of operations produces the same answers, there is little to marvel at (Spinoza, 1677): if you carry out the same assay you will produce the same object.[24]

By this introduction to the microprocessess of the fact production, we have tried to show that a close inspection of laboratory life provides a useful means of tackling problems usually taken up by epistemologists; that the analysis of these microprocesses does not in any way require the a priori acceptance of any special character of scientific activity; and finally that it is important to eschew arguments about the external reality and outside efficacy of scientific products to account for the stabilisation of facts, because such reality and efficacy are the consequence rather than the cause of scientific activity.

NOTES

1. In this chapter we use only a small fraction of the material bearing upon microprocesses. Our intention is merely to provide an overview of the work of the laboratory. To do this we have had somewhat to simplify the analysis of conversations and accounts. A full analysis, particularly one that aspired to the rigour of "conversational analysis" (for example, Sacks, 1972; Sacks et al., 1974), would demand much more detailed treatment than given here.

2. The problem of indexicality in science has already enjoyed limited attention. Barnes and Law (1976), for example, have argued that none of the expressions used by scientists can escape indexicality. The implication is that scientific expressions are no better able to yield a determinacy of meaning than any employed in "nonscientific" or common sense contexts. Garfinkel's (1967) discussion can also be read as supporting this conclusion. In a related manner, a number of continental semioticians have recently begun to extend the tools of literary analysis to the study of rhetoric in a wide number of areas: poetry, advertisements, lawyers' pleas, and science (Greimas, 1976; Bastide, forthcoming; Latour and Fabbri, 1977). For semioticians, science is a form of fiction or discourse like any other (Foucault, 1966), one effect of which is the "truth effect," which (like all other literary effects) arises from textual characteristics, such as the tense of verbs, the structure of enunciation, modalities, and so on. Despite the enormous difference between Anglo-Saxon studies of the ways in which indexicality is repaired and continental semiotics, they hold in common the position that scientific discourse has no privileged status. Science is characterised neither by an ability to escape indexicality, nor by an absence of rhetorical or persuasive devices.

3. This phenomenon was observed many times in the study. It does not imply that papers are prejudiced or that there is widespread tampering with data. Rather it demonstrates, as we suggested in Chapter 2, that papers are operations in a field which are loaded so as to make the operations more effective. The relation between data and points is analogous to the relation between ammunition and targets. This is why there is no reason why papers should be an accurate reflection of the research activity of the laboratory (Medawar, 1964; Knorr, forthcoming).

4. To see others' comments as dangerous objections depends, in turn, on Smith's career decisions. If he left science (and went into teaching) his sensitivity to objections might be modified. By contrast, we showed in Chapter 3 how objections could be taken as very serious, even though they turned out to have no significance.

5. Such technical discussions are not intrinsically different from others; they correspond to a certain stage and pressures within the agonistic field. Wilson's transition from questions of theory ("how would you explain this mechanism?") to questions of general technique ("on which assay do you try that?") depended on his confidence in his colleagues. When his confidence was really low, he would ask more specific questions ("show me your book"), and if it did not go well at all, Wilson on some occasions probed the use of relatively menial procedures ("which sample did you use, where did you take the powder? How did you number the racks?") His confidence and vested interests were crucial to the kind of questions in which he engaged.

6. In most of the discussions of the past, issues of the correct allocation of credit were the main focus. See Chapter 5.

7. One main advantage of our anthropological perspective is its reliance on a wealth of written documents: papers, protocol books, articles in journals, letters, and

even the transcripts of conversations. As long as such written documents are available, the tools of semiotics, exegesis, and ethnomethodology can be applied. At first sight, however, "thought processes" do not lend themselves to this kind of treatment.

8. This operation matches Hesse's (1966) definition of analogical process. In terms of a sorting process, the special interest of X for cancer is sorted *out*, the notion of a superimposition between selenium content in water and *something* which varies is sorted *in* and imported into Slovik's specific problem. The analogical resemblance which explains the proximity and the subsequent step, is the phenomenon which varies from one region to the other. Our interest is not the analogical reasoning per se but the absence of such reasoning (analogical or otherwise).

9. The notion of ideas as a summarised account which enhances belief in the existence of a thinking self, owes much to Nietzsche's (1974a; 1974b) treatment of scientific truth.

10. The simple transformation of statements about things, into stories specific to a particular genre, is the basis of the *Formgeschichte* (Bultmann, 1921). Although obvious when dealing with biblical exegesis,this transformation has not enjoyed much attention in the study of science.

11. Crick and Watson (1977) have since explained how Watson's confidence in Donohue became strong enough to overcome his belief in the authoritative chemistry textbook. The participants recalled that the fact that Donohue was the only person (apart from Pauling) who could be believed was crucial.

12. Once again, this example matches Hesse's (1966) scheme. Bis's work on neurotensin is sorted out, the principle of a temperature assay is borrowed and imported into the bombesin area. The link which makes the connection possible is the similarity between bombesin and neurotensin. However, the crossing or hybridisation concerns physical events rather than notions or concepts: an assay is crossed with a substance.

13. The expression is taken from the referee's report: "The findings *per se* are an expansion of the orginal work of B and his co-workers on neurotensin, but the marked potency of bombesin . . . on temperature is an *outstanding* contribution." The terms extension and outstanding indicate the referee's grasp of the analogical process. The first published paper retains some traces of the analogical path: "Because of the similarities of the biological activities of these peptides and their distribution in the CNS, we have tested several naturally occurring peptides." The subsequent paper starts from a new role of these peptides in the central nervous system.

14. Of course, the adoption of this perspective was a practical necessity. The participants were themselves very much aware that they were engaged in construction.

15. This has been the stock in trade of philosophers since Hume's radical treatment of the problem.

16. When asked to describe the object of a statement which has been "discovered," scientists invariably repeated the statement. By repeating the same statement in less detail, however, it is possible to convey the impression that there is *more* to reality than is being said. The incompleteness of this description is taken as an indication that the object is not entirely exhausted by knowledge of it (See Sartre, 1943).

17. The history of the construction of this substance will be related in detail elsewhere. By contrast with the case of TRF, the observer was present from initial attempts to construct this substance up until its final solidification and use in industrial processes.

18. The question now raised is what kind of explanation is applicable to the settlement of controversy, given that its truth statement cannot be used. Although we

indicate some of the answers in the case of TRF, and go on to outline a general model of explanation in Chapter 6, our main intention here is to extricate the question from the remnants of the realist position.

19. Our use of the term shadow contrasts with Plato's original use of the term. For us, reality (ideas in Plato's terms) is the shadow of scientific practice.

20. Frequently, in histories of epistemology (for example, Bachelard, 1934) the argument of efficacy is used when the argument of truth becomes untenable; conventionalists take over (Poincaré, 1905) when realists are defeated (and vice versa). The argument that it works is not more nor less mysterious than the argument that it fits reality.

21. Many of the substances (and their analogues) mentioned in earlier chapters are patented. Substances "discovered" in the laboratory are described in the texts of patents as having been "invented." This shows that the ontological status of statements is rarely likely to be finally settled: depending on the prevailing interests of the parties concerned, the "same" substance can be given a new status.

22. The notion of reliability is itself subject to negotiation (Collins, 1974; Bloor, 1976). When, for example, several laboratories failed to confirm the results produced by members of our laboratory, the latter simply recast these failures as evidence of the others' imcompetence (VII, 12).

23. We do not wish to argue another version of the induction problem in philosophical terms; we simply want to put the problem on an empirical footing so as to make it amenable to study by sociologists of science. On an empirical basis, neither TRF nor somatostatin escape the material and social networks in which they are continually constructed and deconstructed. For a discussion of the case of somastatin, see Brazeau and Guillemin (1974).

24. This wonderment is particularly marked in matters of science. Nobody wonders that the first steam engine from Newcastle has now developed into a worldwide railway network. Similarly, nobody takes this *extension* as the proof that a steam engine can circulate even where there are no rails! By the same token, it has to be remembered that the extension of a network is an expensive operation, and that steam engines circulate only on the lines upon which it has made been made to circulate. Even so, observers of science frequently marvel at the "verification" of a fact within a network in which it was constructed. At the same time they happily forget the cost of the extension of the network. The only explanation for this double standard is that a fact is supposed to be an idea. Unfortunately, empirical observation of laboratories make this idealisation of facts impossible.

Chapter 5

CYCLES OF CREDIT

Each of the preceding chapters has portrayed laboratory life from a somewhat different perspective. The anthropological approach of Chapter 2 demonstrated the importance of literary inscription in the laboratory; the historical treatment in Chapter 3 showed the dependence of facts on their construction within a particular material context; and Chapter 4 encroached on the ground of epistemology in order to demonstrate the microprocesses at work in the constitution of phenomena such as "having ideas," "using logical arguments," and constructing "proofs." One advantage of this style of presentation is that, for the most part, we have been able to cut across many of the distinctions with which the study of science is often associated. For example, in Chapter 3 we were able to analyse scientific activity without commitment to either side of the distinction between fact and artefact. Similarly, in the last chapter, we attempted to examine the operation of microprocesses without committing ourselves to either a realist or relativist position. The main reason for our not wanting to ally ourselves with one or the other side of these distinctions is that we found that these distinctions provided a resource for participants in the laboratory. It seems inappropriate to use such distinctions in order to understand laboratory activity when they were themselves found to be constituted through this activity.

One particular distinction has remained unexplicated in our discussion so far, although it has been hinted at during various stages of the argument. We refer to the distinction between the production of facts and the individuals involved in this production. Of course, we have already referred to the workforce responsible for the activation of inscription devices (Ch. 2), the makers of decisions and investments (Ch. 3) and the proponents of ideas and arguments (Ch. 4). Nevertheless, we have as yet said little about scientists as individuals. In particular, we have avoided taking the individual scientist as the starting point or main unit of analysis in our discussion. This may seem odd in an essay avowedly concerned with the *social* construction of facts. Yet it fits well with our observations of laboratory life: the overall impression which emerges from the field notes is that before being an individual or a mind, each of our informants was part of a laboratory. Consequently, it was the work sequences, networks, and techniques of argument which suggested themselves as more appropriate units of analysis than individuals. In addition, we noticed that the distinction between the individual and the work done by him provided an important resource in the construction of facts. Our informants were constantly engaged in debates concerning the place of the individual and his or her association with work which had been done. As we noted earlier (Ch. 1), the invocation of the presence of human agency can provide a powerful means of undercutting claims to facticity. On several occasions, informants reported that it was they who had had a certain idea; subsequently, however, other members of the laboratory reported the same idea to have resulted from "the group's thinking process." The observation that the distinction between individuals and their activity acted as a resource for participants was a further reason for our unwillingness to take the individual as the starting point for analysis.

In this chapter we examine the currency of this distinction and look at the way it was persuasively employed in the laboratory. Many of the scientists we observed had successfully used the distinction to construct an individual's career for themselves, a career which was quite clearly separated from the material and economic aspects of laboratory activity. Less successful participants, such as some of the technicians, found themselves with careers which were inextricably bound up with the material elements of the laboratory. We shall attempt to account for the construction of individual careers without separating the resulting individual from the activity of fact construction in the course of which he is created. To do this, we use the notion of credit to link together aspects of laboratory activity which are usually discussed under the rubrics of sociology, economics, and epistemology. In the first part of the chapter, we argue that an extended notion of credit can link together these apparently disparate aspects of laboratory activity; in the second part of the chapter, we apply this notion of credit to the careers and group structure of our particular laboratory.[1]

Credit: Reward and Credibility

WHAT MOTIVATES SCIENTISTS?

What drives scientists to set up inscription devices, write papers, construct objects, and occupy different positions? What makes a scientist migrate from one subject to another, from one laboratory to another, to choose this or that method, this or that piece of data, this or that stylistic form, this or that analogical path? One way to answer these questions is to postulate norms impressed upon the scientist during his training and silently enforced during his subsequent career. As has been noted elsewhere, however, attempts to derive the existence of norms from the kind of material available to us are prone to major difficulties (Mulkay, 1975). In particular, we were unable to identify explicit appeal to the norms of science, except in very few instances. Some of these more nearly constituted an appeal to counternorms (Mitroff, 1974): "everyone pushes ones' own stuff, it's normal—Normal?—I mean human" (IV, 57). Other remarks seemed designed solely to give a good impression. For example, when asking his technician to set up an instrument for the next bioassay, Nathan said: "If we don't do that double check, people could argue that the numbers in our paper are due to. . . . " When later asked why he had used this instrument, Nathan replied: "In science, you should always be overcautious" (X,2). The justification in terms of possible debate and criticism was thus recast for the benefit of the outside observer in terms of norms. Of course, it could be said on the basis of Nathan's last statement that norms are present but invisible, but even if we grant this kind of inferential leap, norms could not explain the choices of laboratory, subject area, or an item of data. At best, norms simply delineate large-scale trends in behaviour; at worst, they simply refer to themes of ideological discourse (Mulkay, 1975). In either case, the explanatory power of norms falls well short of our objective of understanding both science and the scientists who make it.

An alternative approach to the explanation of scientist's behaviour pays more careful attention to the terms in which they themselves explain their behaviour. While appeals to norms were extremely rare among our respondents, the description of activity in quasi-economic terms was pervasive, especially among younger scientists.[2] Consider the following examples:

This instrument can bring me ten papers a year (II, 95).

We had a sort of joint account with him; he got the credit, we got it too; now we cannot draw on it anymore (VI, 12).

Why working on this (substance), we are not the best in this area; we invested a lot in the releasing factor field . . . we are the best in it, we'd better stay in it (VII, 183).

Here, then, are typical instances of the use of notions of investment and return. Such usage was not limited to occasional utterances; sometimes it was sustained throughout long and more sophisticated explanations of career patterns. In one exchange, for example, A volunteered an overall picture of why people do science. His explanation was a complex mixture of liberal political economics, social darwinism, cybernetics, and endocrinology:

[A]ll depends on feedback, what is your threshold of satisfaction, what is the quality, and the frequency of the feedback you need . . . it is difficult to handle all the variables. I was a physician . . . I wanted a place where you are paid more than $20,000 a year . . . this was obligatory medicine . . . but I wanted positive feedback proving my smartness . . . patients are not so good for that . . . I want a very rare commodity: recognition from peers . . . I moved to science . . . O.K., but I am a high achiever . . . I do not need frequent feedback like Bradt so I can choose subjects which are not too rewarding at the beginning (VI, 52).

Most widely shared were the evaluations by a new investigator of opportunity in his field. On five occasions during interviews, respondents sketched a curve representing the growth of their discipline and explained why they entered it or left depending on fluctuations in the curve. For example:

[T]his is peptide chemistry, see . . . it's tapering off. . . . I knew Brunick's laboratory works only on that, so I didn't go there, but now . . . [draws another ascending curve] this is the future, molecular biology, and I knew that this lab. would move faster to this new area (XIII, 30).

Whether or not these statements correspond more to respondent's real motives than to a set of convenient justifications, we are unable to resolve. However, we take as significant respondents' constant reference to investment, rewarding studies and exciting opportunities. Frequently, they related their efforts to what they called market fluctuations and drew lines to show how these fluctuations explained their behaviour. The complexity of these self-representations through economic or business metaphors contrasts sharply with the simplicity of norms. This complexity is exemplified by T's explanation for his wish to leave science and go into teaching:

Science has been for me largely dissatisfying by comparison with the investment I put in it . . . I can anticipate that it will be like that later . . . I really worked harder than I thought justified for the amount of positive feedback that I got. . . .
Q: What do you mean by that?
T: By positive feedback I mean the satisfaction of having a problem solved and the gratification obtained in communicating it to others (VI, 71).

T went on to explain that he regretted leaving science, but that it was a case of all or nothing especially because his work was "not cheap research. . . . I'd need $100,000 at least, to equip a laboratory." On the other hand, he thought it possible that the state to which he was moving might become rich enough to be able to fund him. He added:

[M]y ability to find a job in research again will be increased in one year when the papers we are writing now will be published . . . but if I wait one more year after teaching, I will be definitely screwed up (VI, 73).

T's calculation thus involved consideration of available funds, the extent of positive feedback, the general funding policy of a particular state, and the publication and reception of his papers. Since all these factors were reckoned to vary with time, T's major concern was to determine when best to capitalise on the opportunities available.

Of course, the frequent use by informants of economic analogies does not mean that economic models are necessarily the best explanation of their behaviour. It does suggest, however, that explanations exclusively in terms of social norms are inadequate. More significantly, it is clear from the examples above that scientists talked about data, policy and their careers almost in the same breath. They thus appeared to be working with a model of their own behaviour which made no distinction between internal and external factors.

LIMITATIONS OF THE NOTION OF CREDIT AS REWARD

A possible interpretation of the above examples is that scientists are using economic metaphors to talk about credit. It might be said, for example, that assessments of opportunity and return from investment are metaphorical reformulations of the processes whereby credit is allocated. It is true that a good deal of laboratory conversations included mention of the term credit. The observers' notebooks reveal the almost daily reference to the distribution of credit. Furthermore, the term credit was used explicitly by informants in interview. Overall, credit was used in four ways. Firstly, it was a commodity which could be exchanged. For example, the end of a letter of thanks for the loan of some slides included the following:

Thank you again for giving me the opportunity to use them in future lectures. Please rest assured that I will, of course, credit you for them.

Secondly, credit could be shared:

[H]e shared most of the credit with me, which was very generous of him, because I was a young puppy at the time.

Thirdly, it could be stolen:

He says *my* laboratory, but it is not his, it's our's and we are going to do all the work, but he will get all the credit.

Fourthly, it could be either accumulated or wasted. These different usages indicate that credit has all the character of a currency. As we shall show, however, overreliance on explanations of scientists' behaviour in terms of their quest for this currency, entails misleading oversimplification.

The prevalence of references to credit made us suspicious. An outsider, especially one bearing the label of sociologist, might well expect to be fed stories about credit because this is regarded as suitable material for exponents of what is often perceived as an essentially muckraking endeavour. Since respondents are unable, at least initially, to discuss the details of their scientific work with outsiders, they tend to respond in terms of topics which they feel are appropriate, that is, gossip, scandal, and rumour. Consequently, we would expect reference to credit to be more frequent in exchanges with outside observers than with other participants. In our laboratory, this effect was exacerbated by the prevalence of strong feelings about certain recent instances of the maldistribution of credit. On many occasions respondents had to be persuaded to discuss the process of science rather than the allocation of credit! Clearly, certain local conditions accounted for the unusual predominance of reference to credit.[3]

Although scientists discussed credit, they did not do so all the time. In particular, little mention was made of credit when they discussed their data or when they talked about the future. When asked in interviews why they had come to this laboratory, or why they had chosen a particular problem area or method, not one of our twenty interviewees answered in terms of the availability of credit. Here then is a paradox: participants talked freely and even tirelessly about credit in some situations but never mentioned it in others. By looking carefully at these two sets of situations, one gets the impression that although important, the notion of credit as reward is a secondary phenomenon. For example, it was only at the end of a long letter asking for substances, proposing experiments, and suggesting ideas that Herbert offered his thanks for the reception at a recent meeting and added: "About your earlier work with . . . you certainly deserve all the credit for these early and astute behavioural observations." It is, however, not possible to account for the rest of the letter on the basis of this reference to a past episode. At the end of a discussion with A, C commented: "You will get a lot of credit for that." But this scarcely permits us to explain their whole two-hour discussion in terms of their quest for credit. At the end of his long report, one referee wrote: "Dopamine was first reported to inhibit . . . in vitro by Mc . . . [ref] who should be quoted here." The referee could be seen to be invoking a rule of credit sharing at this point. But this does not explain the wealth of his previous comments. References to credit can frequently be found, but it only assumes prominence in discussions of the past, or of group structure, or of issues of priority. Consequently, credit as reward cannot adequately account for the behaviour of a scientist practising science. Rather, it explains a limited set of phenomena, such as the delayed repartition of resources in the aftermath of some scientific achievement.

It is, of course, possible to argue that scientists are motivated by a quest for credit even though they do not talk about it and deny that credit in the form of reward is their motive. But this would require the existence of a system of repression to explain how the real motive (credit) never appears consciously in the participant's account of his motivation. Instead of pursuing ad hoc explanations it may be better to suppose that scientists are not motivated simply by credit. If, for example, informants report in interviews that they chose a certain method because it yielded reliable data, is their reference to reliability to be taken as a disguised form of concern about obtaining credit? When another respondent reports that he wanted to solve the problem of how the learning process works at the brain level, is this to be understood as obscure way of saying that he wanted credit?

THE QUEST FOR CREDIBILITY

The Oxford Dictionary gives several definitions of credit, only one of which ("recognition of merit") corresponds to the sense in which some sociologists use the term to denote credit as reward. The alternative dictionary definitions are:

(1) The attribute of being generally believed in . . . credibility.
(2) Personal influence based on the confidence of others.
(3) Reputation of solvency and probity in business, enabling a person or body to be trusted with goods or money in expectation of future payment.

It is clear, therefore, that credit can also be associated with belief, power, and business activity. For our laboratory scientists, credit had a much wider sense than simple reference to reward. In particular, their use of credit suggests an integrated economic model of the production of facts. In order to examine this possibility, let us look in some detail at the career of one scientist and assess which definition of credit most usefully explains it.

In interview, Dietrich revealed that after obtaining a medical degree he moved out of medicine in order to do research: "I was not very interested in money, research was more interesting more difficult and challenging" (XI, 85). His next decision was where to do graduate studies: "Bern was not bad, but Munich was a much better place, more prestigious and more interesting" (XI, 85). As has been shown elsewhere, the location of a scientist's training has a significant influence on his or her future career. In economic terms, graduate training at Munich was worth several times more than the same training at Bern. In other words, Dietrich realised that he would be more highly accredited if he trained at Munich. From this we can see that the beginning of a scientist's career entails a series of decisions by which individuals gradually accumulate a stock of credentials. These credentials correspond to the evaluation by others of possible future investments in Dietrich.

Then I went to a Congress at Eilat . . . I realized the interest of neurophysiology. . . . This seemed to be a good field, not crowded, bound to be more and more important . . . not like cancer which one day will be solved and put to an end (XI, 85).

Dietrich thus explained his decision to work in neurophysiology in terms of his interest. At the same time, we can see the elements of a quasi-economical calculation through which a young investigator evaluated the opportunities of a field and his chances in it. The evaluation of his prospects entailed an assessment of the likely return from his own investment of effort. Dietrich's next step was to choose someone working in the field.

I heard about X at this Congress. I went to see him, but he turned me down . . . he did not want physicians . . . he didn't want to form a group of young guys . . . it's a waste of time (XI, 85).

According to what he had heard at the meeting, Dietrich knew that X was the best in the field. For Dietrich, this meant that the same investment in X's group would be much more effective than in any other. The hiring process entailed negotiation during which either side attempted to evaluate the capital that the other could offer.

But X told me to see Y at [Institute]. . . . Y told me to work on that subject, that it should be finished in a year, and that he would support me in getting tenure at . . . the subject was to localize an enzyme in the brain . . . he was completely wrong about the timing because it is still an open question . . . but I wanted a position so I followed his advice. . . . I got a position at. . . . I wrote my dissertation and had several publications (XI, 85).

This is a good example of a smooth start to a career. The inscription devices worked and sufficient documents were generated to support his papers and dissertation. In short, Y's investment had paid off. But the return in terms of reward was marginal. Dietrich's work was neither widely acclaimed nor regarded as an outstanding achievement. Yet with Y's backing it was enough to secure him tenure. Dietrich was now an accredited researcher who was able to work in the field in earnest.

This enzyme was not well studied before. I showed that what people said before was wrong . . . they purified it 1000 times and claimed it was pure, I purified it 30,000 times and showed that it was still not pure. . . . I can say that I advanced toward the characterization of this enzyme (XI, 85).

This contribution represented an incremental scientific advance, with all the elements of a typical operation (see Chapter 2), a change in standards of purification and a parallel change in technique. Dietrich could sum up his position thus: "Curiously, a lot of people studied the degradation of acetylcholine but very few the synthesis. . . . I am the world expert [laughing] . . . on this enzyme." This particular producer of facts established access to a market for his contributions. As a result, he would be invited to any meeting which discussed this enzyme. He would be cited in any paper dealing with this issue. He was thus able to transform his small savings into greater revenue.

To do the mapping in the brain with fluorescent methods you need an antibody which is monospecific, but to raise this antibody you need a pure enzyme. For me, I told you, even purified 30,000 times, it is still not pure enough to be specific . . . but someone in Houston claimed he has a pure enzyme.

In order to obtain credible data he required a particular inscription-device with specific technical capabilities. Clearly, if too much noise was generated, the data could not be warranted as reliable. In the market there was a demand for a pure enzyme; since it was not information which could be communicated, Dietrich had to move to Houston in order to collaborate with Z. Dietrich hoped to obtain new data by using his own methods on Z's pure material. The project was a failure, however, because Z's claim was not supported by any data. Z did not have the enzyme. But Dietrich had access to other more important resources and saw his opportunity within another specialty.

I have always been interested in peptides. . . . I was blocked a bit in this, my boss was an impossible guy . . . also I knew Parine and I wanted to go to the West Coast.

Dietrich was able to obtain money in the form of a fellowship to work with Flower at the institute. Fellowships constitute an advance made by private or federal institutions to investigators once they have proven their solvency. Subsequently, such advances are refunded indirectly by way of publication and facts. "At least, I had shown I could work by myself, that's the most important thing."

By chance, Parine put Dietrich to work on a subject of considerably greater importance than his previous enzyme study. In other words, the same amount of work had a much greater impact in the new field (in terms of access to funds, reception of citations and invitation to congresses than it would have done in the first. As a result of his association with S, Dietrich received increasingly attractive offers (in terms of space, technicians, independence, and materials) to persuade him to return to Germany. "You see, I am now a specialist in peptides at a time when it's ripe in Germany, and when they have very few of them" (XI, 86). In the Institute, Dietrich enjoyed greater access to a much more active market than had been the case in Germany. The simple fact of his association with S and W gave him substantial credibility, both in terms of prestige and material resources. By being at the institute, Dietrich obtained access to communication networks, substances, and technicians, and he was able to tap the vast capital of material resources described in Chapter 2. The investments made by Dietrich had an enormous payoff both because of a concentration of credit in the institute and because of a high demand for credible information in the field. In addition, his German nationality enabled him to play on the variations between currencies. He could obtain a much higher return for his efforts in Germany as a result of his work in the U.S. But the laboratory space, the technicians, the independence, and grant money made available to him in Germany were not offered as a form of reward. Rather, these were material resources to be quickly reinvested in new inscription devices, and in the production of data, papers, and facts. Were these investments in his work not to pay off, Dietrich would lose credibility. In this respect, scientists' behaviour is remarkably similar to that of an investor of capital. An accumulation of credibility is prerequisite to investment. The greater this stockpile, the more able the investor to reap substantial returns and thus add further to his growing capital.[4]

To repeat, it would be wrong to regard the receipt of reward as the ultimate objective of scientific activity. In fact, the receipt of reward is just one small portion of a large cycle of credibility investment. The essential feature of this cycle is the gain of credibility which enables reinvestment and the further gain of credibility. Consequently, there is no ultimate objective to scientific investment other than the continual redeployment of accumulated resources. It is in this sense that we liken scientists' credibility to a cycle of capital investment.

CONVERSION FROM ONE FORM OF CREDIBILITY TO ANOTHER

Although Dietrich's career path undoubtedly involved a series of decisions based on precise and complex calculations of interest, the exact nature of this interest remains at issue. If we limit ourselves to the notion of the pursuit of reward for scientific contributions, it is clear that Dietrich is bankrupt. After investing for ten years, he is almost unknown, having received less than eight citations a year, no awards, and having made few friends. If, however, we extend the notion of credit to include credibility, we can see a much more successful career. He has good credentials, he has produced credible data using two sorts of methods, and now works in a new and important area at an institution with an enormous accumulation of resources. In terms of his pursuit of reward, his career makes little sense; as an investor of credibility it has been very successful.

By distinguishing between credit as reward and credit as credibility, we are not merely playing on words. Credit as reward refers to the sharing of rewards and awards which symbolise peers' recognition of a past scientific achievement. Credibility on the other hand, concerns scientists' abilities actually to do science. We saw at the end of Chapter 2 how a statement could be transformed from a claim to a fact by the use of documents which made unnecessary the continued inclusion of modalities. Statements thus supported by the appropriate documents can be said to be credible in the same way as individuals are credible or instruments are reliable. The notion of credibility can thus apply both to the very substance of scientific production (facts) and to the influence of external factors, such as money and institutions. The notion of credibility allows the sociologist to relate external factors to internal factors and vice versa. The same notion of credibility can be applied to scientists' investment strategies, to epistemological theories, to the scientific reward system, and to scientific education. Credibility thus allows the sociologist to move without difficulty between these different aspects of social relations in science.

If we suppose that scientists are investors of credibility rather than just reward seekers, we can easily explain a number of otherwise strange cases of scientific behaviour in terms of the conversion by scientists of one form of credibility into another. We can best elucidate this point by way of four examples:

(a) When I consider all the investment I made in this substance in the laboratory and I don't even have a good assay for it: If Ray is unable to set up this assay, he will be fired (XIII, 83).

The investment referred to here was in terms of both money and time. On the basis of this investment, a payoff was expected in the form of data which could support an argument in a forthcoming article. The worth of the person put in charge of the assay depended on the quality of the assay and the data produced. If the assay failed, Ray would lose credibility and would lose both his investment and the data required to support his argument. Consequently, X warned Ray (albeit indirectly) that his position was at stake. In this case, data from the bioassay was necessary to support an argument. Success in the bioassay was necessary to support Ray's authority. This authority was necessary, in turn, to support his position. Lastly, X's investments had to be supported or repaid by a new paper.

(b) The peak of the field has passed . . . it really boomed after P's experiment like that. . . . A lot of people flooded into that field and . . . after a while, when nothing new happened it seemed more and more impossible . . . . Expectations were so high that people published papers without any experiment, just speculation. . . . Then a lot of people got negative answers when they tried to replicate that . . . the accumulation of negative results dampened expectations (VIII, 37).

As a result, a number of people, including P, began to leave the field. The initial experiment had prompted a small gold-rush and career paths had changed direction as people invested in a new field. Initially, standards were such that no experiment was necessary. Almost any proposition was accredited in the prevailing atmosphere of excitement. When hard data began to flow, however, a large variety of propositions were bankrupted one after the other. Negative results thus modified career expectations once again.

Speaking about an investigator in another field, Y said:

(c) I supported the earlier results of this guy . . . when a lot of people took that as garbage; he is a big shot in his field . . . so now he invited me to meetings and it's a good occasion for me to meet new people in another field (X, 48).

Y's faith in another scientists's proposition was eventually converted into invitations to meetings. Furthermore, this invitation provided a good occasion to get acquainted with others and to become informed of new ideas. The same information would subsequently be converted into a new experiment. Thus, confidence in someone else's data, which were thought controversial, constituted a capital investment. The investment could be repaid in this example by virtue of the other scientist's position ("he is a big shot").

K and L were counting samples on the beta counter. K is fifteen years older than L.

(d) L: Look at these figures, it's not bad.
K: Well, believe in my experience, when it's not much more above 100, it's not good, it's noise.
L: The noise is pretty consistent though.
K: It does not change much, but with this noise you can't convince people . . . I mean good people (XIII, 32).

From the perspective of some epistemologists, we would expect the reliability of data to be an issue quite distinctly separated from the evaluation of individuals in the field. Thus, the assessment of data should not be so obviously linked to the rhetorical operation of convincing others and should vary neither according to the individual who is doing the interpreting nor according to the audience to whom the results are addressed. Nevertheless, examples such as the above reveal that scientists frequently make connections between these superficially foreign issues. In fact, such issues are all part of one cycle of credibility. Consequently, the connections made between them can be explained in terms of the conversion between different forms of credibility. It is not surprising, therefore, that a participant simultaneously evalues the quality of data, the standing of the audience, and his own career strategy.[5]

Figure 5.1 illustrates the cycle of credibility. The notion of credibility makes possible the conversion between money, data, prestige, credentials, problem areas, argument, papers, and so on. Whereas many studies of science focus on one or other small section of this circle, our argument is that each facet is but one part of an endless cycle of investment and conversion. If, for example, we portray scientists as motivated by a search for reward, only a small minority of the observed activity can be explained. If instead we suppose that scientists are engaged in a quest for credibility, we are better able to

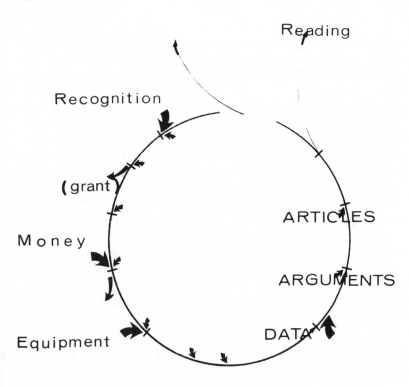

Figure 5.1
This figure represents the conversion between one type of capital and another which is necessary for a scientist to make a move in the scientific field. The diagram shows that the complete circle is the object of the present anaysis, rather than any one particular section. As with monetary capital, the size and speed of conversion is the major criterion by which the efficiency of an operation is established. It should be noted that terms corresponding to different approaches (for example, economic and epistemological), are united in the phases of a single cycle.

make sense both of their different interests and of the process by which one kind of credit is transformed into another.[6]

THE DEMAND FOR CREDIBLE INFORMATION

In order to understand the full force of the difference between reward and credibility, it is necessary to distinguish between the process by which reward is bestowed and the process by which credibility is assessed. Both reward and credibility originate essentially from peers' comments about other scientists. Thus, even the award of a Nobel Prize depends on various submissions, recommendations, and assessments by working scientists. But what form do these evaluative comments take in the laboratory itself? Two features are readily apparent. Firstly, evaluative comments made by scientists make no distinction between scientists as people and their scientific claims. Secondly, the main thrust of these comments turn on an assessment of the credibility which can be invested in an individual's claim. The possibility of bestowing reward is a marginal consideration. A striking illustration of this is provided by the following example: C and Parine were in the bioassay room when C asked Glenn to synthesize a peptide which another colleague, T, had claimed to be more active than endorphin. When a syringe of the peptide had been prepared, C made ready to inject a rat on the surgery table:

I bet you the peptide is going to do nothing . . . this is the confidence I have in my friend T. [C squeezed the syringe and enjoined the rat]: O.K., Charles T., tell us. [A few minutes passed.] See, nothing happened . . . if anything the rat is even stiffer [sigh]. Ah, my friend T . . . I went to his laboratory in New York and saw his records . . . which lead to publication . . . it made me feel uncomfortable (V, 53).

This incident underscores the common conflation of colleague and his substance: the credibility of the proposal and of the proposer are identical. If the substance had the desired effect on the rat, T's credibility would have increased. If, on the other hand, he had had more confidence in T, C would have been surprised by the result. This is made particularly clear in the following:

Last week, my prestige was very low, X said I was not reliable, that my results were poor and that he was not impressed. . . . Yesterday, I showed him my results . . . good lord, now he is very nice, he says, he was very impressed and that I will get a lot of credit for that (XI, 85).

For a working scientist, the most vital question is not "Did I repay my debt in the form of recognition because of the good paper he wrote?" but "Is he reliable enough to be believed? Can I trust him/his claim? Is he going to provide me with hard facts?" Scientists are thus interested in one another not because they are forced by a special system of norms to acknowledge others' achievements, but because each needs the other in order to increase his own production of credible information.

Our discussion of the demand for credible information contrasts with two influential models of the exchange system in science proposed by Hagstrom (1965) and Bourdieu (1975b). Both models have obviously been influenced by economics. Hagstrom's model employs the economics of preindustrial societies and portrays the relation between two scientists as that of gift exchange. According to Hagstrom, however, the expectation of exchange is never made explicit:

> [T]he public disavowal of the expectation of recognition in return for scientific contributions should no more be taken to mean that the expectation is absent than the magnanimous front of the kula trader can be taken to mean that he does not expect a return gift (Hagstrom, 1965: 14).

Explicit reference to the expectation of exchange occurred in many of the cases we observed. There was no suggestion that our scientists had to maintain a fiction that they were not expecting any return gift. Consequently, the basic argument that scientists are gift givers does not seem warranted. Indeed, we can pose the same question which Hagstrom himself asked:

> But why should gift giving be important in science when it is essentially obsolete as a form of exchange in most other areas of modern life, especially the most distinctly "civilized" areas? (Hagstrom, 1965: 19)

Hagstrom provides no reasons for the survival of this antiquated tradition in the scientific community other than the fact that the same phenomenon is evident in other professional spheres. In all such professional spheres, Hagstrom argues,

> The gift exchange [or the norm of service], as opposed to barter or contractual exchange is particularly well suited to social systems in which great reliance is placed on the ability of well-socialized persons to operate independently of formal controls (Hagstrom, 1965: 21).

For Hagstrom, then, the archaic system of gift exchange is functionally corequisite to the maintenance of social norms. In other words, the archaic system of potlach is seen as a way of supporting the central system of norms. Even scientists' publication strategies are manifestations of conformity to norms through participation in gift exchange:

> The desire to obtain social recognition induces the scientist to conform to scientific norms by contributing his discoveries to the larger community (Hagstrom, 1965: 16).

Scientific activity is governed by norms, and the enforcement of these norms entails the existence of a special system of gift giving. But this system is never made mention of by participants. Indeed, if scientists deny they are expecting a return gift, this can be taken as proof of the success of their training and their rigorous conformity to the norms. We have here an explanation of an exchange system in terms of norms which is both empirically undemonstrable and which the author himself considers an inexplicable and paradoxical archaism.

Why should Hagstrom use a primitive exchange analogy to explain relationships between scientists? We had the distinct impression that the constant investment and transformation of credibility taking place in the laboratory mirrored economic operations typical of modern capitalism. Hagstrom was struck by the apparent absence of transfers of money. But this feature should not lead to the formulation of a model designed to preserve the existence of norms. Do scientists read each other out of deference to norms? Does one individual read a paper so as to force its author to read his work in return? Hagstrom's exchange system has the aura of a rather contrived fairy tale: scientists read papers as a matter of courtesy, and similarly thank their authors out of politeness. Let us look at one more example of scientific exchange in order to show that this view is needlessly complicated.

One of the main problems in studying diabetes was the difficulty of discriminating between the effects of insulin and glucagon in a diabetic patient's glucose level. In other words, attempts to study the effects of insulin were foiled because of the "noise" generated by glucagon, the effects of which it was impossible to suppress. In 1974, however, a new substance called somatostatin was isolated (in a completely unrelated field), which was found to inhibit the secretion of both growth hormones and glucagon (Brazeau and Guillemin, 1974). Somatostatin was immediately imported into the field of diabetic study and used to decrease the effect of glucagon.

> The discovery of GH releasing inhibiting hormone, Somatostatin, might open the way to an objective evaluation of the role of glucagon in

diabetes. It will soon be possible to follow diabetic patients with competely suppressed glucagon secretion.

This passage, written by a clinician, indicates the potential importance of glucagon. If at this point, someone had told the clinicians that he knew the structure of a glucagon suppressor substance, he would have been seized violently by the lapels. Why? Because the clinician would have felt overcome by a desire to reward this individual for his contribution? Or because he felt a debt of honour to the individual's achievement? *No.* The clinician's violent reaction would stem from his ability, once armed with the new information, to rush to his bench or his ward and set up a protocol in which one of the causes of noise in his inscription device can be controlled. The clinician would not be obliged to disburse credit to the bearer of information, nor even to cite his paper. The utility of the information for the generation of fresh information is crucial, whereas the subsequent bestowal of recognition is only a secondary concern to the scientist.

Bourdieu's model of scientific exchange compares scientists' behaviour with that of modern businessmen rather than precapitalist dealers and traders. The absence of money in scientific exchange does not cut any ice with him because of his experience in studying exchange systems in fields other than science. For Bourdieu (1975b), economic exchange can include the accumulation and investment of resources other than money. By using the idea of symbolic capital Bourdieu describes the investment strategies in fields such as education or art in terms of modern capitalism. Even business strategies are analyzed from the point of view of accumulation of symbolic (rather than just monetary) capital. By contrast with Hagstrom, Bourdieu (1975b) does not attempt to explain scientists' behaviour in terms of norms. Norms, the socialisation processes, deviance, and reward are the consequences of social activity rather than its causes. Similarly, Bourdieu takes the position that science can be studied without forging ad hoc explanations and in terms of other more usual rules of economics. For Bourdieu, then, the cause of social activity is the set of strategies adopted by investors wanting to maximise their symbolic profit.

The scientific field is the locus of a competitive struggle, in which the specific issue at stake is the monopoly of scientific authority, defined inseparably as technical capacity and social power (Bourdieu, 1975b: 19).

Investors' strategies are likened to any other businessman's strategy. However, it is not made clear why scientists should be interested in one another's production. Bourdieu simply asserts:

> [T]he transmutation of the anarchic antagonism of particular interests into a scientific dialectic becomes more and more complete as the interest that each producer of symbolic goods has in producing products that, as Fred Reif puts it "are not only interesting to himself but also important to others" . . . comes up against competitors more capable of applying the same means (Bourdieu, 1975b: 33).

This tautological explanation of interest is worsened by the absence of any reference to the content of the science produced. In particular, there is no analysis of the way in which technical capacity is linked to social power. This absence might not be a problem in the study of "haute couture" (Bourdieu, 1975a), but it is absurd in science.

Neither Bourdieu nor Hagstrom helps us understand why scientists have any interest in reading each other. Their use of economic models, derived respectively from capitalist and precapitalist economies, fails to consider *demand*. This failure corresponds to their failure to deal with the contents of the science. As Callon (1975) has argued, economic models can be applied only if this accounts for the content of science. Hagstrom and Bourdieu provides useful explanations of the repartition of credit as a sharing process but they contribute little to an understanding of the *production of value*.

Let us suppose that scientists are investors of credibility. The result is the creation of a *market*. Information now has value because, as we saw above, it allows other investigators to produce information which facilitates the return of invested capital. There is a *demand* from investors for information which may increase the power of their own inscription devices, and there is a *supply* of information from other investors. The forces of supply and demand create the *value* of the commodity, which fluctuates constantly depending on supply, demand, the number of investigators, and the equipment of the producers. Taking into account the fluctuation of this market, scientists invest their credibility where it is likely to be most rewarding. Their assessment of these fluctuations both explains scientists' reference to "interesting problems," "rewarding subjects," "good methods," and "reliable colleagues" and explains why scientists constantly move between problem areas, entering into new collaborative projects, grasping and dropping hypotheses as the circumstances demand, shifting between one method and another and submitting everything to the goal of extending the credibility cycle.[7]

It would be mistaken to take the central feature of our market model as the simple exchange of goods for currency. Indeed, at the preliminary stage of fact production, the straightforward exchange of information for reward is hindered by the fact that the scientist and the claim are *not* distinguished. What then is the equivalent of buying in our economic model of scientific activity? Our scientists only rarely assessed the success of their operations in terms of formal credit. For example, they had little idea of the extent to which their work was cited. They were not normally concerned about the distribution of awards, and they were only marginally interested in questions of credit and priority.[8] Indeed, our scientists had a much more subtle way of accounting success than simply measuring returns in currency. The success of each investment was evaluated in terms of the extent to which it facilitated the rapid conversion of credibility and the scientist's progression through the cycle. For example, a successful investment might mean that people phone him, his abstracts are accepted, others show interest in his work, he is believed more easily and listened to with greater attention, he is offered better positions, his assays work well, data flow more reliably and form a more credible picture. The objective of market activity is to extend and *speed up the credibility cycle as a whole*. Those unfamiliar with daily scientific activity will find this portrayal of scientific activity strange unless they realise that only rarely is information itself "bought." Rather, the object of "purchase" is the scientist's ability to produce some sort of information in the future. The relationship between scientists is more like that between small corporations than that between a grocer and his customer. Corporations measure their success by looking at the growth of their operations and the intensity of the circulation of capital.[9]

Before using this model to interpret the behaviour of our laboratory scientists, it is important to stress its complete independence of any argument concerning motivations. Explanations using the notion of reward required us to suppose that scientists routinely hide their real motivations when they fail to reveal an explicit interest in credit and recognition. By contrast, our credibility model can accommodate a variety of types of motivations. It is not necessary, therefore, to doubt the motivations expressed in informants' accounts. Scientists are thus free to report interest in solving difficult problems, in getting tenure, in wanting to alleviate the miseries of humanity, in manipulating scientific instruments, or even in the pursuit of true knowledge. Differences in the expression of motivation are matters of psychological make-up, ideological climate, group pressure, fashion, and so on.[10] Since the credibility cycle is one single circle through which one form of credit can be converted into another, it makes no difference whether scientists variously insist on the primacy of credible data, credentials, or funding as their prime motivating influence. No matter which section of the cycle they choose to emphasize or consider as the objective of investment, they will necessarily have to go through all the other sections as well.

Strategies, Positions, and Career Trajectories

In the first part of this chapter we discussed scientists' investments and portrayed them as investors of credibility. We shall now attempt to apply the notion of credibility to the particular situation of our laboratory scientists.

CURRICULUM VITAE

A scientist's curriculum vitae (C.V.) represents a balance sheet of all his or her investments to date. A typical C.V. contains name, age, gender, family information, and four sections, each of which corresponds to a particular meaning of credibility. Under "Education," for example, we may read:

1962: Bachelor of Science and Agriculture, Vancouver
1964: Master of Science, Vancouver, B.C., Canada
1968: Ph.D. (Cellular Biology), University of California

This list of qualifications represents what could be called the scientist's *accreditation*. This in itself does not ensure that the individual *is* a scientist, but it does enable him to be admitted to the game. In investment terms, this individual has the necessary credentials to invest. Such credentials represent the formal return on a large loan of taxpayers' money (or sometimes private funds) invested in education and training. Of course, the date, location, and subject matter of each qualification are all important. For example,

Dr. Hoagland holds his bachelor's degree from Columbia, his master's from MIT and his Ph.D. from Harvard (Meiter et al., 1975: 145).

These qualifications are understood to be more impressive than those in the previous example (Reif, 1961). Similarly, if the subject matter of a scientist's doctoral examination includes bacterial genetics, he has a distinct advantage when applying to collaborate with a group which requires expertise in this area. A scientist's qualifications constitute cultural capital which is the successful outcome of multiple investments in terms of time, money, energy, and ability. The scientists and technicians in our laboratory had accumulated more than one hundred thirty years of college and graduate education.

Qualifications such as the Ph.D. do not differentiate between scientists because virtually all scientists possess one. More important is the information contained in the second section entitled "Positions."

1970   Assistant Research Professor, The Institute
1968-70   Postdoctoral Research Chemist, University of
                California, Riverside
1967-68   Fellow Research Assistant, University of
                California, Riverside

This information indicates both that an individual has been admitted to the game and that he has actually played sufficiently well to have obtained a position. For the same reason, C.V.'s record any grants and awards which have been received:

(1) Alpha Omega Alpha, Hoover Medical Society, Alpha of
     Arizona Chapter
(2) Prizer Scholar
(3) Arizona Medical Student Research Award
     Public Health Service Trainee in Endocrinology, 1965 to 1969
     Public Health Services Postdoctoral Fellowship

The list of grants and awards provides a statement of the extent of investment already placed in the individual. Thus, the statement of an individual's credibility represented by his qualifications and position is reinforced. One further source of reinforcement is the inclusion of

the names of advisors and head of laboratories where an individual has
worked:

1973-75   Visting research chemist, laboratory of Nathan
           O. Hakan, Department of Chemistry, University of
           Haifa
1966-68   Postdoctoral Fellow, Microbiology Institute,
           University of Copenhagen, Denmark, N. O. Kierkegaard,
           sponsor

The inclusion of these names, together with those of referees from
whom letters of recommendation can be requested, reflect the importance of established relationships as a source of credibility. Readers
can use these names both to ascertain the network in which a scientist
is situated and to identify sources who can vouch for his or her
solvency.

Of course, none of these characteristics of curricula vitae are
peculiar to researchers. What is particular is not so much a scientist's
academic position (or appointment) but rather his position in the field.
Readers may wish to know which problems the scientist has solved,
which set of techniques and expertise he or she is familiar with and
which problems the scientist is likely to be able to solve in the future.
Frequently, however, the statement of academic positions and positions in the field are conflated:

Positions
1962-64   Synthesis of pyrrole compounds, State College
1964-65   Conduct freshman chemistry laboratory, Stanford University
1965-69   Isolation and structure elucidation of alkaloids,
           Stanford University
1969-70   X-ray crystallography, Stanford University
1970-   Research Associate, The Institute

The first four of these positions concern problems undertaken at a
particularly prestigious location, the last is the academic position
eventually obtained through the conversion of previously accumulated
credibility.

Lists of publications are the main indicators of the strategical
positions occupied by a scientist. Names of coauthors, titles of
articles, journals in which they have been published, and the size of the

list together determine the scientist's total *value*. Once a C.V. has been read and letters of recommendation received, a decision is made on the basis of an individual's value whether or not to give tenure, to grant money, to hire or simply to collaborate on a particular research programme. The C.V. can thus be compared to the annual budget report of a corporation.

The previously accumulated capital of laboratory members was small because they had published relatively little before becoming part of the laboratory group. Eleven scientists had published only sixty-seven items between them and half of these resulted from the work of one individual who had already left the laboratory by the end of our study. In addition, members of the laboratory had held few academic positions before coming to the laboratory. All but one had previously been postdoctoral fellows. In terms of capital, therefore, members of the laboratory provided more the promise of credibility than an accumulated stockpile.

POSITIONS

Scientists move from one position to another attempting to occupy what they regard as the best possible position. It is important to note, however, that each position simultaneously comprises academic rank (such as postdoctoral fellow or tenured professor), a situation in the field (the nature of the problem being tackled and the methods used), and geographical location (the particular laboratory and the identity of colleagues). This three-fold notion of position is crucial to an understanding of scientists' careers. If the analyst does not simultaneously take these three aspects into account he is liable to produce either a conceptual representation of the field (where problems generate other problems), or a picture of individuals struggling against the forces of administration, or a structure of political economy which focusses on institutions, budgets, and science policies. But the cohesion of these three aspects will escape his attention.

The field[11] does not appear full of more or less interesting problems but for the presence of an individual with ambitions to make some points. Nevertheless, the individual strategy is nothing but what the field forces require. The notion of position is thus very complex. It points to the intersection of individual strategy and field configuration, but neither the field nor the individual are independent variables. Let us consider an analogy with war in order to elucidate this point.[12]

A small earth mound is of no obvious strategical significance in itself. If, however, a battle takes place in the vicinity, then this mound *may* take on a special significance. Although at one time merely part of a landscape, it is potentially a *strategical position*. But it only takes on such significance by virtue of a strategist's evaluation of the battlefield, the positions of other troops and the relative strength of the combatants. To one of the combatants, this mound may appear to provide the opportunity for a successful attack on one line of the enemy. The mound suddenly makes sense. He becomes excited by what he regards as an extraordinary opportunity and begins to mobilize the forces at his disposal. He anticipates that once the mound is transformed in position he will be able to effect devastating moves against the enemy. Consequently, he tries to reach and occupy it. The success of his endeavour depends on the state of play in the rest of the battlefield, on the strength of his own forces and on his skill in command and in the evaluation of danger. Once he has reached his objective and transformed the innocent mound into a *point d'appui*, the pressures of the battlefield will immediately be modified. Others may try to force him out. His ability to withstand these pressures depends, once again, on his past ability, the resources at his disposal (men, weapons, and munitions), the resources provided by the mound (better visibility, dominant situation, rocks, etc.), and his skill in using them. A position is similarly the resultant of a participant's career trajectory, the situation in the field, the resources at his command and the advantages of the invested position.

The above analogy fits closely to scientists' strategies as revealed in interviews. Scientific activity in our laboratory comprised a field of contention in which facts were produced, claims dissolved, artefacts deconstructed, proofs and arguments disproved, careers ruined, and prestige cut down. This field only existed in so far as it was perceived by participants. Furthermore, the precise nature of this perception depended on participants' initial standing. Again and again we were told: "Then I got interested in this technique, this area, this guy" or "I realised the interest of" or "I saw an opportunity" and so on. Respondents described how they seized a specific method or inscription device and brought it to a particular place where they began to make points and to publish. Repeatedly, we heard in interviews that "it did not work" or that a respondent "was getting nowhere." Respondents related how they then drifted until they found an instrument, a method, a collaborator or an idea that worked. They were then able quickly to modify the situation in the field. Some statements which they discredited were never taken up by others. They became strong. They gained weight. They obtained more funds, attracted more assistants, generated arguments. *The field was modified around their new position.*

The strategical concept of scientific activity is exemplified by Guillemin's experience in the field of releasing factors. When he first entered the field, Guillemin perceived a central problem to be that of obtaining a reliable bioassay for TRF. Having decided on a strategy, he mobilized colleagues in the pursuit of such an assay and grasped the chance opportunity of assistance from a lady whose skill perfectly matched his goal. He quickly began to obtain reliable data on the basis of which he shot down a number of existing claims and postulated the existence of TRF, for which he immediately gained the recognition of others. Similarly, Dietrich was prevented from mapping the brain because of the absence of an antibody, the production of which depended on the isolation of a pure enzyme. As a result he decided to move to a country to collaborate with researchers who possessed the enzyme. His move depended almost entirely on the position in which he wanted to invest.

It becomes clear that sociological elements such as status, rank, award, past accreditation, and social situation are merely resources utilised in the struggle for credible information and increased credibility. It is at best misleading to argue that scientists are engaged, on the one hand, in the rational production of hard science and, on the other, in political calculation of assets and investments. On the contrary, they are strategists, choosing the most opportune moment, engaging in potentially fruitful collaborations, evaluating and grasping opportunities, and rushing to credited information. In interviews it is not merely peripheral concerns which excite and interest them. Their political ability is invested in the heart of doing science. The better politicians and strategists they are, the better the science they produce.

It is important to realise, however, that our definition of position is purely relative. In other words, a position has no meaning without a field or set of participants' strategies. At the same time, the field itself is no more than the ensemble of positions as evaluated by a participant. Moreover, a participant's strategy is meaningless unless located within a field and in relation to positions as perceived by other participants.[13] The notion of position should not be reified. A position does not exist "out there," simply waiting for someone to fill it, even though this is how it appears to the actor. Indeed, the nature of positions to be seized is constantly the focus of negotiation in the field. The feeling that a constraint on taking a position is dependent on the field is also the upshot of constant negotiation. Positions are only retrospectively defined as being available for occupation. But this kind of perception is again only relative to the field in the sense that when we say "G occupied a position," this is shorthand for our retrospective understanding of the way in which G determined the configuration of the field, his resources, and his career. The scientist himself may retrospectively justify his occupancy of the position in terms of its interest.[14]

TRAJECTORIES

The rather monotonous pattern of participants' remarks about career strategies is a reflection of the monotony of the investment process:

I studied this problem. I met Dr. Maddox, I developed this technique, I published this paper, then a position was offered at this place, I met Sweetzer, we published this paper. I decided to move to this area.

Participants' careers comprise a number of successively occupied positions. Moves from one position to another can be evaluated by devising a kind of balance sheet which presents individual careers in terms of the credit (cultural capital, social capital, operations) with which they started and the positions in which they invested. The perceived success of each move and the crude index of impact used in Chapter 2 (number of citations per paper published after each move) are also recorded. Each row of the balance sheet thus represents one move, that is, a change in position (Table 5.1). An individual can thus move to another laboratory with the same subject and academic status, or he can stay in the laboratory but change his problem area, or he can change academic rank without modifying his research programme. Participants start each move with an initial capital together with their earnings from previous moves. Since capital can be wasted, individuals' accounts can sometimes go into the red. For example, Sparrow joined the laboratory with a Ph.D. in biochemistry and letters of recommendation. These credentials were no better than average. However, Sparrow's first paper turned out to be an extraordinarily good investment. He synthesized a releasing factor and received

Table 5.1

	Academic Position	Position in the Field	Geographic Position	Payoff
	None	None	Bern	Medical Doctorate
	Graduate Student	None	Munich	Training
1968				
	=	Neurophysiology	X's LAB	=
1970				
	Tenure	Enzyme purification	=	Ph.D. and tenure
1972				
	=	Isolation of the enzyme	=	expert, invited to meetings
1973				
	=	=	U.S. Z'lab—Houston	=
1975				
	=	Brain peptides	California— Flower's LAB	=
1976				
	=	=	=	Known everywhere because of work with Flower and C. on brain peptides
1978				
	Full professor	=	Germany—head of laboratory	=

Table 5.1 represents a simplified balance sheet of Dietrich's moves. Each row corresponds to a move, in which one of the three aspects of Dietrich's position was modified. Each column corresponds to Dietrich's career trajectory as measured in terms of one aspect of his position. The right-hand column records the pay-off resulting from each move. An equals sign (=) indicates that no change took place.

hundreds of citations, largely because the releasing factor related to particularly sensitive areas of medicine (such as sterility) and because its synthesis had important implications for birth control. In other words, a large audience required the use of the newly synthesized substance in hundreds of experiments. His six coauthors lent him part of their capital (in the form of instruments, expertise, space, and credibility) in such a way that his own contribution was difficult to distinguish. He remained in the same area for four years and continued to synthesize analogues of the same substance but his efforts met with diminishing returns. (Up to 1976, he received 0, 0, 10, 4, 3, 2 and 0 citations for each of his subsequent 7 papers.) He then decided to move to another problem area in order to work on his own. But he did not realise that most of his capital had resulted from his location and the demand for the specific releasing factor he had synthesized. As a result, he suddenly found himself without access to space in the institute or grant money and with no more personal credibility than when he started. His attempt to change position corresponded to a failure to convert his accumulated credibility because the credibility was not properly his to convert. Subsequently, he was fired by the institute and tried to exchange his scientific capital for a teaching position or industrial work in chemistry. This entailed his giving up the chance of obtaining any further scientific credibility. His movement out of the credibility cycle amounted to the liquidation of his scientific investments.

The importance of location is well illustrated by the trajectory of scientists who entered the laboratory at the beginning of their careers and who left after a short while. A comparison of productivity, measured by number of citations per paper in the three years following publication, of five scientists reveals marked differences between the period before, during and after their stay in the laboratory (Table 5.2) Although all five clearly benefitted from their research in the laboratory, four of them were unable to reinvest, or cash, their acquired credibility once they moved away. One obtained a better research position but did not publish anything which has since been cited and three others had to liquidate their assets either by teaching, or by going into business. In terms of credibility, of course, these moves represent poor invest-ments. In terms of money or security, however, there may well have been a significant payoff. The last of the five obtained a tenured position in research, partly because he already possessed his own independent capital. This was sufficient, when taken together with his stay in the laboratory, to be exchanged for tenure: "There is no doubt that it helped me tremendously" (IV, 98).

GROUP STRUCTURE

From the point of view of the production of facts, a group can be thought of as the result of the intertwining of several trajectories. Group organisation can thus be interpreted in terms of the accumu-lated moves and investments of its members. The conjunction of participants' trajectories make up a hierarchy of administrative positions. Our laboratory group formed an almost perfect administra-

Table 5.2

Scientist	Before	During	After	Conversion
G	0	13	0	business
S	0	8	0	teaching, business
F	2.5	36.6	0	better research position
U	0	10	0	industry
V	14	22	-	better research position

tive pyramid. A wide base of fifteen unqualified technicians was headed by five senior technicians who, in turn, were responsible to eight professional researchers (Ph.D. holders). These eight comprised five assistant research professors, two associate research professors, and one full professor (who was also director).[15]

The sociological functions corresponding to these administrative positions related directly to the part played by each individual in the process of fact production. We saw in Chapter 2 that the field of releasing factors is both capital and labour intensive. Thus, information was obtained from a bio- or radio-immunoassay, which typically occupied several individuals for weeks at a time. We saw in Chapter 3 how some of the difficulties of this kind of work were met by the accumulation in one place of a large workforce, body of skills, and equipment. Part of the work was automated by labour-saving machines, such as the automatic pipette and automatic counters. For the most part, the technicians were responsible for this work, which provided data to be used in the arguments of the scientists.

A technician's status depends on the extent or range of operations with which he is concerned. Thus, the status of technicians whose job is merely to wash glassware is significantly lower than those of jobs entailing responsibility for a complete process, such as the Edmann degradation method for peptide sequencing, or for an entire inscription device, such as the Nuclear Magnetic Resonance Spectrometer or a radioimmunoassay (see Ch. 2). At intermediary status levels, technicians specialise in one or more routine tasks, such as the care of animals or pipetting.

This distinction is not always very clear, however, particularly in cases where technicians assume some of the responsibilities of the scientists. Bran, for example, a technician whose name appears on published papers, commented:

I know more about isolation chemistry than X (a scientist)

[When asked why he was about to leave the group, Bran replied]: I am blocked here, I guess . . . Yes, I love research. I really love it, and it's why I chose to come here . . . but I am blocked. I don't have the ability to get a Ph.D.

Q: The ability or the possibility?

A: No the ability . . . to do research you need imagination, originality. . . . I cannot reach that level . . . it's extremely crowded, and how could I get a Ph.D. here these days . . . it's not the money. I am better payed than Y. . . . Also, I guess I don't want to become a super-tech . . . yes, you know somebody with a Ph.D. but who does not do any mental work. . . . I could see more than a few who are super-techs, here . . . maybe its the IQ, I don't have the IQ for doing research; I don't want to struggle for years for getting a Ph.D. and after that just being a super-tech (IV, 88).

Unlike scientists, technicians did not normally possess the initial capital of credibility (a Ph.D.) which could be used to gain further credibility. Although technicians were less interested in cashing and reinvesting scientific credibility than in a salary, they showed an intense concern for the distribution of credit and for the wording of acknowledgments. In economic terms, technicians were more akin to workers rather than investors. Their salary repaid their labour, but it did not constitute capital which could be invested. This is not to deny their use of various strategies to better their positions, for example, by moving to another laboratory. But such moves could never secure parity with investors who possessed a Ph.D. This is why no less than five young technicians left the laboratory during our study in order to follow courses towards a Ph.D. With this initial qualification, technicians expected their work to bring both a salary and an increment in credibility which could be further invested.[16]

Bran saw "super-techs" as qualified scientists who simply carried out routine work for others. Indeed, he argued that a Ph.D. would have been of little use since many of the scientists with Ph.D.s spent most of their time doing the work of technicians. For Bran, the difference between a technician and a "super-tech" was insufficient to justify an investment of several years hard work. What then characterises the supertechnician Ph.D. holders?

The citation histories of the eight scientists in the laboratory are markedly different. Three scientists received an average of 150 citations a year and the remainder received about 50 citations a year. This difference, between what have been termed "major and minor leaguers" (Cole and Cole, 1973), is even more striking when we look at the citation spectra of individuals' publications (Figures 5.2a, 5.2b, and 5.2c). Each spectrum reveals the extent of citation for each paper cited more than twice in one year. Citation spectra thus indicate the span of a participant's career, the repartition of effort and success and the obsolescence of each paper. For example, F's spectrum (not shown) indicates that he received citations for only one paper. A, on the other hand, had a healthy spectrum (not shown) even though he received a relatively small total number of citations. This difference exemplifies the difference between leaders (major leaguers) and the supertechnicians (minor leaguers). On average, minor leaguers were better paid than technicians and they tended to be first authors on papers. These papers were cited, but this small amount of credibility was insufficient to provide the authors with resources, such as independent space or grant money. Thus, minor leaguers made points in the literature as well as producing data. But the production of data usually resulted from decisions taken by major leaguers. Minor leaguers set up complicated bioassays, synthesized peptides and collaborated with others when asked to do so. This provided them with opportunities to write a paper. but the main moves were made by those on whose initiative the peptide bioassay was set up or the collaboration was effected in the first place. Between 1970 and 1975 the 4 major leaguers wrote 100 papers as first authors and received 8.3 citations per paper in following years, while the 8 minor leaguers wrote only 70 papers and received seven citations per paper.[17]

Another key feature of the hierarchy is the extent to which people are regarded as replaceable. Since the value of information is thought to depend on its originality, the higher a participant in the hierarchy the less replaceable he is thought to be. Supertechnicians are seen as less easily replaceable than good technicians, who are seen as less easy to replace than routine workers. But the glasswasher and the gardener can be changed without affecting the process of fact making. For example, one of the major leaguers commented on the imminent departure from the laboratory of one of the supertechnicians in the following way: "of course, we will use a synthetic chemist of some sort."

According to this respondent, another individual could fulfill the function of providing substances just as efficiently as the departing chemist. At the same time, this respondent regarded her own work quite differently; but for her presence less new information would have been produced.[18] It is difficult to account for the careers of the eight minor leaguers by saying that their investment in a field has been effective, because supertechnicians work mainly for others and tend

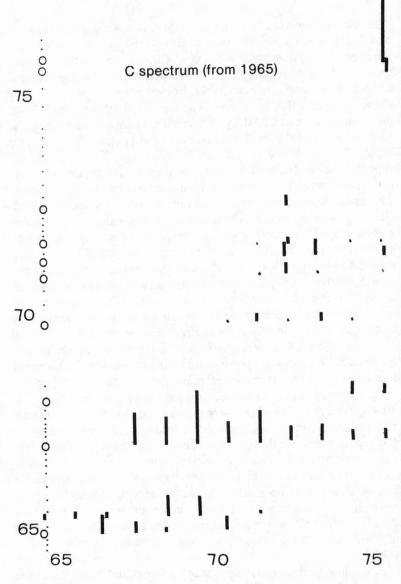

Figure 5.2a
The extended reception of a scientist's work can be illustrated by a "spectrum," which draws both on the number of items published by a scientist (as first author) and also on the impact of these items in terms of received citations. Articles are represented at the time of their pub-

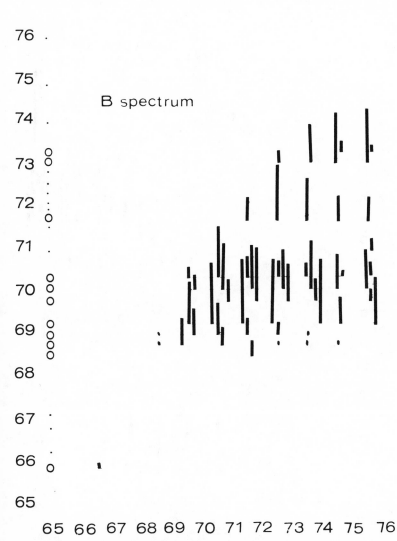

Figure 5.2b

lication by a point on the vertical time scale; if they are subsequently cited more than twice, articles are represented by a circle. The citation history of each article (source: SCI) is denoted by vertical bars proportional to the number of citations received in a given year (horizontal time scale). The spectrum thus provides a graphic summary of individual

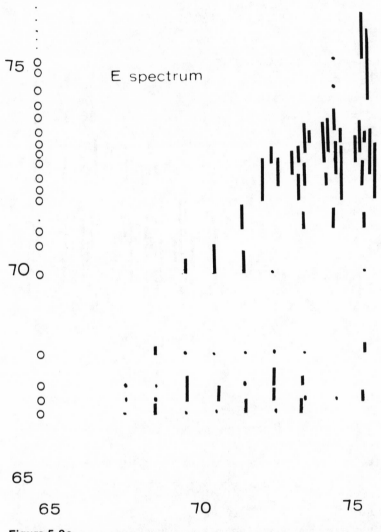

Figure 5.2c

scientist's careers. C (Fig. 5.2a) can be seen to have enjoyed relatively little success in articles published between 1967 and 1975. B's spectrum (Fig. 5.2b) reveals rapid aging in that his more recent publications have attracted little attention. By contrast, E (Fig. 5.2c) has a healthy spectrum, with nearly all his recent publications enjoying citation.

not to effect substantial gains in capital. On the contrary, they are unable to buy positions or grant money. They can, however, lend their skill to an investigator, in exchange for a secure position and some nonmaterial satisfactions. They thus circulate on the market in a similar way to senior technicians. They are hired, not because of their originality, but on the recommendation of an investigator for their reliability in producing certain types of data which are needed by another investigator in order to make new points.

The leaders of the laboratory have to create original information. One of them, the chairman, can hire technicians, and scientists to work under him. He has sufficient capital of credibility to make unnecessary its direct reinvestment in bench work. He is a capitalist par excellence, since he can see his capital increase substantially without having directly to engage in the work himself. His work is that of full-time investor. Instead of producing data and making points, he tries to ensure that research is pursued in potentially rewarding areas, that credible data are produced, that the laboratory receives the largest possible share of credit, money and collaboration and that conversions from one type of credibility to another can occur as swiftly as possible.

GROUP DYNAMICS

In order to understand the dynamics of the group we have to look at the history of its investments, as reconstructed from curriculum vitae and interviews. Occasionally, when an anthropologist is lucky enough to witness the disintegration of a tribe and the subsequent creation of a new settlement, he can catch a glimpse of those rules of behaviour that remain hidden during periods of normal activity. By chance, our study of the laboratory coincided with the negotiation of a completely new research contract and with the disbandment of the group. Before turning to this, however, let us look briefly at the way in which the group had developed before the time of our study.

Between 1952 and 1969, C accumulated a large capital of credibility by occupying a unique position—the releasing factor area. This position hinged on his suggestion of methods which were still in use some twenty-five years later, and on his imposition of a certain set of rigorous standards (Ch. 3). On the basis of this, he was elected to the Academy of Science, received a series of progressively larger grants and managed to persuade a chemist (B) with a good career already behind him to join the group. At the same time, C trained two young students, who subsequently became his pre- and then post-doctoral fellows. The collaboration between C and B paid off in 1969, with the solution to a structure. This brought the group immense credit. C also invested substantial effort in the isolation of another substance with implications for the problem of birth control. At this point the possibility was raised of setting up a completely new laboratory with three times more personnel and what was described as "the best equipment in the world." The potential application of the results of the kind of research being conducted by C, together with his existing credibility and the group's success, all made possible a new settlement at the institute.

Between 1969 and 1972, the number of citations received by the group increased. As a result of his work in chemistry, B received substantial credit and became head of a new laboratory with a team of three senior chemists. E benefited both from working in a large physiology group and from the experience as informal leader of a team of two (and later three) investigators. His work on the mode of actions and analogs of newly characterised substances increased his standing in the field. The whole group was organised like an assembly line producing a series of new structures. The structure of somatostatin became a fresh source of credibility for the group because, by chance, its synthesis was found to have important implications for the treatment of diabetes. Whereas C received a number of awards and lecture invitations for this work, B and E obtained what they considered a more important kind of return: credibility. Although C did little bench work he devoted considerable energy to the exchange of work done by the others for grant money so as to maintain or enlarge production activity in the laboratory. Thus, the relationship between C and the others constituted a kind of "joint account." As C increasingly became the nationally known figurehead of the group, he did less and less work on his own, and the number of citations to his papers decreased (see Fig. 5.3).

Between 1972 and 1975, the lack of success in producing a new substance was accompanied by changes in the internal structure of the group. Several scientists left for opportunities elsewhere. B, for example, saw his access to work in chemistry limited by the concentration of his skills in one particular research programme. His aptitude to produce information decreased, as did the number of citations he received. Unable to renew his capital, he began to see his position weaken and his status lowered, even though his academic position remained unchanged. Two of the young supertechnicians, H and G, adapted easily to the routine of the second research programme (the

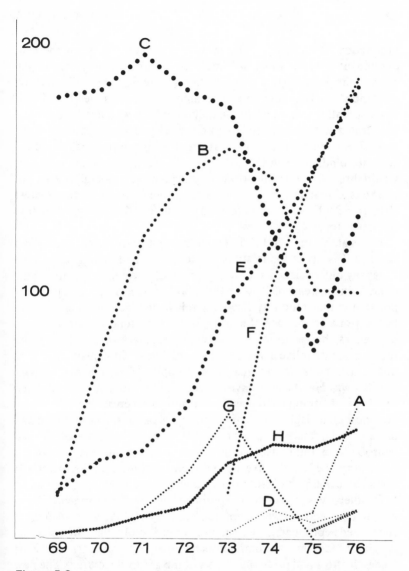

Figure 5.3
The SCI was used to determine the total number of citations received
by each member of the group per year, starting in 1969 when the group
took on its present shape. Unlike Figure 5.2, this computation does
not take into account which article is cited. Nevertheless, a comparison of the curves gives a rough approximation of the weight of the
scientists. The crossing of the curves in different years corresponds

production of analogs). They took over responsibility for the production of analogs while still maintaining an auxiliary role in the work of the physiology section. With his increased credibility, E took over the physiology section and became to be seen as the official boss of its operations. A multimillion dollar contract with a federal agency was drawn up to guarantee laboratory work in diabetes, birth control, and CNS effects for five years. It was C's signature which made the contract binding, even though there was tacit understanding that E would direct the scientific work. At this point, C's capital (in terms of citations as first author) was at low ebb while E had by far the greatest (Figure 5.3). E, A, H, and I formed the nucleus of a new group within the laboratory.

It was at this time, in 1975, that the present study began, largely as a result of C's invitation to study epistemology and biology and to see "the way older scientists leave a group and let younger scientists take over." But instead of leaving the laboratory in order to further his position in the credibility circuit, C reinvested his time and energy in bench work. In the face of many jokes and the total scepticism of his colleagues, he set to work in amongst the glassware, columns and bioassays, rather like a new postdoctoral fellow. Obviously, this work utilised the immense resources of the group. But C carried out the work on his own. He chose to invest three months work in a problem he regarded as strategic: the isolation and characterisation of a new peptide which displays the same activity as opiates. This problem had already been tackled in other fields, such as pharmacology and neurobiology. But C decided that by drawing on the resources of the laboratory, he could solve the problem in three months, using the classical techniques of isolation chemistry and physiology. According to C, other investigations of the problem had been uninformed: "These people do not know what peptide chemistry is." As it happened, he succeeded in producing the structure in a little more than three months, despite the fact that his competitors had spent several years on it. This research effort had profound effects of the group structure.[19] The new substance, which could then be produced in large quantities (by virtue

closely to changes in group structure as revealed by interviews. Especially striking is C's comeback after 1975, the slow elimination of B, E's continuous ascension, and the lingering difference between the "big shots" and the "minor leaguers." However, it is only by combining this diagram with the individual spectra that a full idea of a career can be provided.

of the second research programme, (see pp. 61ff), was of great significance both for pharmacology and brain chemistry, which were then boom areas, and for problems of drug addiction and mental illness. Because these enormous interests were at stake, C's position was completely transformed in the space of six months. In September 1975, he was an old "has been" who wanted to retire. By the following March he was the most sought after member of the group, not because of his past prestige, but because of his fresh credibility in the new field. The dramatic increase in C's citations in 1976 (Fig. 5.3) was entirely due to his new work.

This new move completely disrupted the existing contract by which C obtained the rewards, but the others gained credibility. At the same time, discovery of the new substance established a much stronger link between brain study and endocrinology than had releasing factors, despite the greater interest of the latter for endocrinologists as compared to neurologists. The new substance aroused intense interest among brain scientists, especially those newly established in a nearby laboratory. Thus, on the basis of only a few months' work, C found himself admirably positioned in a new field. B and E, on the other hand, found themselves in a rut. They continued to write papers on classical releasing factors with progressively diminishing returns (Fig. 5.2b and c). C no longer wished to retire, finding himself in a position similar to that at the start of the TRF story.

This example of a sudden change in position highlights the sense in which credit and reward are important to scientists. C invested all his credit as resources in a new area. Largely by means of telephone contact with a number of other laboratories he launched large-scale investigations, exchanged substances, sera, and new data within the newly defined subfield. By virtue of his contacts with Parine (see p. 196) he became a member of an entirely distinct invisible college. Other research efforts in the group were eclipsed by the spectacular success of the new substances. Equipment and technicians were increasingly mobilized to assist in the new task. C and others realised that the entire laboratory capacity could easily be invested in an area potentially much more rewarding than releasing factors. However, A began increasing his investment in a set of new substances of only marginal significance for the main programme, in an attempt quickly to increase his returns. The corporation was breaking up. A new contract had to be defined.[20]

By comparison with the issue of production strategy, factors such as personality or the *"point d'honneur"* played a relatively minor part in the long series of conflicts which accompanied the break up of the group. For five years, the group had existed on the basis of an agreement between three senior investors to work on the same problem at a time when this represented a highly efficient means of subscribing to a given paradigm. With a change in both the field and individual strategies, however, the situation had to be modified. The equipment, money, and authority which made up the dead capital of the laboratory had to be redistributed. B was eliminated and bankrupt. F and A formed a new group with their supertechnicians H and I. Their problem was to decide how and where this new group could settle. The credibility of this new group attracted good offers (chairmanship, lab space, endowment) from several parts of the country even though none of these matched the situation in the laboratory before the success of C's changed strategy. For his part, C felt sufficiently confident in his ability to acquire new capital that he could envisage being left by the group from whom he had obtained his past credit and starting again with a new group of young postdoctorates.

The complexity of the relations between the members of the group and of their appraisal of the definition of credit, became particularly clear once the group had actually split.[21] C was compared to a capitalist in that his full-time activity was to manage his capital and not to work directly to produce credible data. As we have seen, however, his hired hands were also investors in the same market. They could thus become direct competitors of C. This is exactly what happened. E decided to cash his credibility. Quite unexpectedly, he found his credibility sufficient to secure from the same institution a grant enabling him to equip a laboratory exactly similar to the one in which he had been working. He then became head of a group, hired his own staff and secured around him the same equipment that C had before. In economic terms he founded a rival business and employed H, I, A, and most of C's technicians. Figures 5.2c and 5.3 show E's citation curve (together with the newcomer A, and the supertechnician H) to be regularly ascending. B's situation was very different. He was unable to cash any credibility within the field and was forced like Sparrow, to liquidate his assets and go into teaching (Fig. 5.2b). C was left with a large amount of dead capital (in terms of equipment), a little money, but no workforce. He now had to find a new source of points to make in order to activate the mass of former investments incorporated in the laboratory.

The production of credible data, as we have seen, is one way of activating the credibility cycle and of setting in motion the "business of science" or, as Foucault (1978) puts it, "the political economics of truth." Later, scientists may strive to cash credibility in their own name. They may thus say they "have had ideas" (pp. 169ff), that it is "their" laboratory, and that is they who have managed to attract money and equipment in the interests of securing the basis for their operations. From this point of view, they are not unlike businessmen. At the same time, however, they are merely *employees* of the federal government. No matter how extensive, their scientific capital can neither be sold nor bequeathed and only rarely can it be exchanged for monetary capital. As craftsmen working to produce their own data, they are concerned more or less exclusively with their own accounts. But if they are not careful, they can end up as employees or supertechnicians. It is also possible, however, that they can become independent, and with luck, employers themselves. At the same time, they themselves remain employees, in the sense that they are payed to manage the loan of private or taxpayer money which is lent them. The scientists we observed were thus caught between two overlapping economical cycles: they constantly had to manage their capital in order to get things going; but at the same time they had to justify their use of the money and confidence which they had borrowed.

In a successful laboratory there is likely to be constant excitement about finding new statements, proving them, extending their influence, setting up new instruments, cashing credibility, and reinvesting it. The tension of a battalion headquarters at war, or of an executive room in a period of crisis does not compare with the atmosphere of a laboratory on a normal day! This tension is directed towards the secretaries in efforts to persuade them to type manuscripts in time and towards the technicians to effect the rapid order of animals and supplies and to the careful execution of routine assay work. Of course, similar pressures can be found in any production unit. More unusual here is that these pressures force investigators to be credible. On one hand, scientists withstand the constraints of an investor who is continually obliged to reinvest if he does not want to lose his capital. On the other hand, scientists suffer the constraints of an employee constantly required to account for the money lent him. By virtue of this double system of pressures, our scientists remained trapped in the laboratory. If a scientist stopped doing new experiments, occupying new positions, hiring new investigators, and generating new statements, he would very quickly become a "has been." His grant money would be stopped, and, save for any tenured position or niche he had previously established for himself, he would be wiped out of the game. It is possible to explain his behaviour in terms of "norms" or the quest for recognition, but it may not be necessary. Economic forces tie down the researcher both as an independent capitalist and as an employee; in this position it is easy enough to squeeze him so as to extract a fact.[22]

NOTES

1. In this chapter we use loosely structured interviews (many of them tape-recorded), publication lists, curricula vitae, grant proposals, and other documents provided by participants. Valuable data were also obtained through participation in some of the conflicts and dynamics of the group. Our explicit treatment of individual career choices in this chapter has necessitated our taking various precautions, such as changing names, dates, initials, sex and substances on which researchers work in order to protect the anonymity of those concerned.

2. Even in the small group studied here, the representations of the world. or ideologies, differ markedly. Although we did not study these systematically, we payed attention to what Althusser (1974) calls the "spontaneous philosophy of scientists"; one had a typically positivistic representation of science borrowed from Claude Bernard (1865); another had a mystical view of science and linked his work to a fundamentalist approach to religion; a third had a business-like view of his activity and held to the epistemology of a *nouveau riche*; a fourth worked with an economical model of investments; the fifth of the senior members is quoted here.

3. A major problem of this programme is the pressure exerted by informants on the observer to acquire the information they think he wants to hear. This is why we heard so many stories about the politics of the laboratory and why we decided not to use such stories. Behind these stories were very clear strategies of investments, the presence of the observer being used as a resource by which members could determine investments and the nature of others' reactions.

4. Much of this discussion draws heavily on the work of Bourdieu (1972; 1977). The reason for this is simple: economic analyses of science have limited themselves to a consideration of large-scale factors, even when carried out by Marxists like Bernal (1939), Sohn Rethel (1975), and Young (no date). Only by introducing the notion of symbolic capital (of which economic capital is only a subset), is it possible to apply economic arguments to noneconomic behaviour (Bourdieu, 1977). See also Knorr (1978) and Bourdieu (1975b) for a direct application to science.

5. Another example of conversion can be found in Hoagland's reminiscences:

> At Harvard, Gregory Pincus and I had received our Ph.D. degrees in 1927 and had become warm friends. He had remained as an assistant professor in Crozier's department after I left, but after two 3-year terms his appointment was not renewed despite his brilliant work. I was eager to have him join me at Clark

and together we raised sufficient funds from various outside sources to make it possible for him to come as a visiting professor. By 1936 he had published his book, The Eggs of Mammals, and also a number of papers which reported for the first time successful pathogenesis in a mammal, i.e., rabbits that had mothers but no fathers. This received much attention from both the scientific and lay press, but it was met by less than enthusiasm by some conservative members of the University. I found Pincus' interest and knowledge of steroid hormones exciting. He had already developed improved methods of determining urinary steroids and applied these to endocrine problems (Meites et al., 1975).

Each sentence relates the conversion between one form of credibility and another. Thus we read how diplomas, social relationships, positions, money, credit, interests, and convictions interchanged. Hoagland did not simply reward his friend Pincus. Rather, he needed his techniques and his ideas and so backed them and tried to convince others to fund the venture.

6. One main advantage of the notion of cycle, is that it frees us from the necessity of specifying the ultimate motivation behind the social activity which is observed. More precisely, one might suggest that it is the formation of an endless cycle which is responsible for the extraordinary success of science. Marx's (1867: Ch. 4) comments on the sudden conversion from use value to exchange value, could well apply to the scientific production of facts. The reason so many statements are produced is that each is without use value, but has an exchange value which enables conversion and accelerates the reproduction of the credibility cycle. This view also has implications for the so-called relations between science and industry (Latour, 1976).

7. This is typical of the double standard of some analysts of science. When a businessman gives up and sells a bankrupt company, this is taken as an obvious manifestation of greed and interested motives. However, when a scientist gives up a dying area or a discredited hypothesis (which means that no one is going to "buy" the argument any more), this is considered as an indication of conformity to the ethos of scientific disinterestedness.

8. As noted earlier, the laboratory chosen for our study was characterised by an almost pathological concern for credit. It became clear, however, that the *"point d'honneur"* of credit receipt was not itself at stake. Because of the modification of the field, each participant adopted different strategies: the struggle concerned, not credit but space, research programmes, and equipment. As long as they agreed on these points, there was little quarrel about who received credit. When they differed on these points, the tangible focus of conflict was a bitter argument about credit sharing.

9. This comparison is viable in so far as the notion of economics is not restricted to the circulation of money. It should be extended instead to all activities permeated by the existence of a valueless capital, the sole purpose of which is accumulation and expansion. This differs from the efforts of the Chicago School to portray activities in economic terms even where no capital is involved. The link between the scientific production of facts and modern capitalist economics is probably much deeper than a mere relation.

10. A related problem is the extent to which the scientists' activities we portray are conscious and explicit strategies. This is a problem we cannot resolve in abstract because each scientist is also engaged in a debate to make logical, explicit, or necessary his career's choices. We do not wish to say that scientists are "really" interested

although they do now avow it, or that they are "really" determined by the field although they think they have some freedom and merit in having chosen this or that way. We leave questions like the notion of motivations entirely open to psychologists and historians. Some scientists try to show that it was their conscious decision to choose this subject, while simultaneously arguing that a colleague could not do otherwise because the time was ripe. On another occasion the same informant may try to persuade you that he was not conscious at all and that it was some kind of artistic intuition, only to inform you a few days later that the whole thing was quite logical and that he did not have much choice. This consideration is important because we certainly do not wish to propose a model of behaviour in which individuals make calculations in order to maximize their profits. This would be Benthamian economics. The question of the calculation of resources, of maximisation, and of the presence of the individual are so constantly moving that we cannot take them as our points of departure.

11. The word field is used here simultaneously to denote the sense of a scientific field and to convey the idea of an "agonistic field." In this second sense, "field" (Bourdieu's term in French is "champ") denotes the effect on an individual of all others' moves and claims rather than a structure or an organisation. In this way it is not dissimilar to the sense of a magnetic field or to similar uses in physics (magnetic fields, field theory, and so on).

12. Our use of a battlefield analogy is perhaps warranted both by the term field and by the frequent use of military metaphors by scientists themselves (see for example, Ch. 3, p. 130). Although we provide no quantitative evidence, our impressions are that the most frequent use of metaphors in the laboratory was firstly epistemological ("proof," "argument," "convincing," and so on), secondly, economic, thirdly, battlefield analogies, and finally psychological ("pleasure," "efforts," and "passions").

13. As recently argued by Bourdieu at a Paris symposium, the notion of field is understandable only if one takes into account that the nature of motivations, the existence of participants, and the constraints of the field are all themselves at stake in the field. Our argument should in no way be construed as an attempt to resuscitate a structuralist position. An introduction to this debate can be gained from Knorr (1978), Callon (1975), and Latour and Fabbri (1977).

14. In one sense, this entire chapter can be taken as commentary on the frequent utterance of our participants: "That's interesting" (see Davis, 1971).

15. The group of technicians has a high turnover; they are nonunionised and have no long-term contracts; their salaries ranges from $8,000 to $15,000; junior doctorates with no contracts are paid between $12,00 and $20,000; assistant professors with contracts are paid approximately $25,000; associate professors with tenure are paid approximately $40,000. The salary of the head of the group, who has tenure and some power over the space, is unknown. Thus, the salaries are not strikingly different from those in nonscientific companies. More importantly, participants' salaries are insufficient to enable the accumulation of money capital comparable to scientific capital.

16. Seven technicians were interviewed (three tape-recorded) just before they left the laboratory. Their importance in the production of facts is usually underestimated. However, since our main concern is the credibility cycle rather than other more general aspects of laboratory life, we shall not use this interview material here.

17. This difference would be larger but for the generous policy of allowing minor leaguers to assume first authorship.

18. As we mentioned earlier, the struggle for originality is at the heart of fact production. Thus, for participants, the question "How original am I" was the same as "How valuable is my information."

19. Thanks to the SCI (Small, private communication) we are able to confirm that as early as 1977, C was part of a "cluster" with which none of the classical members of neuroendocrinology was associated.

20. This was the situation up until 1977. See below.

21. This is based on a very brief round of follow-up interviews carried out in 1978. The results of events in the recent past is a substantial change in those characteristics of the laboratory described in Chapter 2. Most of the equipment is still there, but only two of the old participants remain. More importantly, although the laboratory was originally fitted out for the production of certain types of fact, it now appears that a rival laboratory is about to flood the market with facts constructed along similar lines. The question for participants is how the equipment described in Chapter 2 can be used in different ways and in different areas. For reasons of space, we are unable to relate these developments in detail. Suffice it to note that the object of our study was a very unusual fit between a group, space, equipment, and a set of problems. The particular situation which allowed us to see many features of fact construction was extremely unusual and may not be repeated.

22. Scientists' final realisation of capital, through their movement into clinical studies, industry, and culture, is not examined here. It is clear, nonetheless, that the sum of investments in the credibility cycle requires eventual justification. This is evident, for example, in scientists' presentation of grant proposals.

Chapter 6

THE CREATION OF ORDER OUT OF DISORDER

In examining the construction of facts in a laboratory, we have presented the general organisation of the setting as constituted by someone unfamiliar with science (Chapter 2); we showed how the history of some of the laboratory's achievements could be used to explain the stabilisation of a "hard" fact (Chapter 3); we then analysed some of the microprocesses by which facts are constructed, looking especially at the paradox of the term fact (Chapter 4); we then turned to the individuals in the laboratory in an attempt to make sense both of their careers and the solidity of their production (Chapter 5). In each of these chapters we defined terms which were often in contradistinction with those used by scientists, historians, epistemologists, and sociologists of science. We shall now summarise the various findings of these preceding chapters in an attempt more systematically to link the different concepts used. At the same time, we shall review some of the methodological problems encountered so far. It will not have escaped the reader's notice, for example, that a major problem arises from our contention that scientific activity comprises the construction and sustenance of fictional accounts which are sometimes transformed into stabilised objects. If this is the case, what is the status of our own constructed account of scientific activity?

In the first section of this chapter we summarise the argument so far. Instead of simply following the presentation of the preceding chapters, however, we identify six main concepts used throughout and show briefly how they are related. This leads us to the second section. Here we introduce one further notion, the concept of order from disorder, which enables us to situate our argument in the more general framework of sociology of science. Finally, in the third section, we compare our own account with those of the scientists whose activity we claim to have understood.

Creating a Laboratory: The Main Elements of Our Argument

The first concept used in our argument is that of *construction*. (Knorr, in press). Construction refers to the slow, practical craftwork by which inscriptions are superimposed and accounts backed up or dismissed. It thus underscores our contention that the difference between object and subject or the difference between facts and artefacts should not be the starting point of the study of scientific activity; rather, it is through practical operations that a statement can be transformed into an object or a fact into an artefact. In the course of Chapter 3, for instance, we followed the collective construction of a chemical structure, and showed how, after eight years of bringing inscription devices to bear on the purified brain extracts, the statement stablilised sufficiently to enable it to switch into another network. It was not simply that TRF was conditioned by social forces, rather it was constructed by and constituted through microsocial phenomena. In Chapter 4, we showed how statements are constantly modalised and demodalised in the course of conversations at the laboratory bench. Argument between scientists transforms some statements into figments of one's subjective imagination, and others into facts of nature. The constant fluctuation of statements' facticity allowed us approximately to describe the different stages in the construction of facts, as if a laboratory was a factory where facts were produced on an assembly line. The demystification of the difference between facts and artefacts was necessary for our discussion (at the end of Chapter 4) of the way in which the term fact can simultaneously mean what is fabricated and what is not fabricated. By observing artefact construction, we showed that reality was the *consequence* of the settlement of a dispute rather that its *cause*. Although obvious, this point has been overlooked by many analysts of science, who have taken the difference between fact and artefact as given and miss the process whereby laboratory scientists strive to *make* it a given.[1]

The second main concept which we have used constantly, is that of *agonistic* (Lyotard, 1975). If facts are constructed through operations designed to effect the dropping of modalities which qualify a given statement, and, more importantly, if reality is the consequence rather than the cause of this construction, this means that a scientist's activity is directed, not toward "reality," but toward these operations on statements. The sum total of these operations is the agonistic field. The notion of agonistic contrasts significantly with the view that scientists are somehow concerned with "nature." Indeed, we have avoided using nature throughout our argument, except in showing that one of its current components, namely the structure of TRF, has been created and incorporated in our view of the body. Nature is a usable concept only as a by-product of agonistic activity.[2] It does not help explain scientists' behaviour. An advantage of the notion of agonistic is that it both incorporates many characteristics of social conflict (such as disputes, forces, and alliance) and explains phenomena hitherto described in epistemological terms (such as proof, fact, and validity). Once it is realised that scientists' actions are oriented toward the agonistic field, there is little to be gained by maintaining the distinction between the "politics" of science and its "truth"; as we showed in Chapters 4 and 5, the same "political" qualities are necessary both to make a point and to out-manoeuvre a competitor.

An agonistic field is in many ways similar to any other political field of contention. Papers are launched which transform statement types. But the many positions which already make up the field influence the likelihood that a given argument will have an effect. An operation may or may not be successful depending on the number of people in the field, the unexpectedness of the point, the personality and institutional attachment of the authors, the stakes,[3] and the style of the paper. This is why scientific fields do not display the orderly pattern with which some analysts of science like to contrast the disorderly tremors of political life. The field of neuroendocrinology thus comprises a multitude of claims and many substances exist only locally. For example, MSH releasing factor exists only in Louisiana, Argentina, and one place in Canada, and in one other in France; most of the associated literature was considered meaningless by our informants.[4] The negotiations as to what counts as a proof or what constitues a good assay are no more or less disorderly than any argument between lawyers or politicians.[5]

Our use of agonistic is not meant to imply any especially wicked or dishonest character attribute of scientists. Although scientists' interaction can appear antagonistic, it is never concerned solely with psychological or personal evaluations of competitors. The solidity of the argument is always central to the dispute. But the constructed character of this solidity means that the agonistic necessarily plays a part in deciding which argument is the more persuasive. Neither agonistic nor construction have been used in our argument as a way of undermining the solidity of scientific facts; the reason for our nonrelativist use of these terms will be clear in our discussion of the third main concept used in our argument.

We have insisted on the importance of the material elements of the laboratory in the production of facts. For instance, in Chapter 2 we demonstrated how the very existence of the objects of study depended on the accumulation inside the laboratory walls of what Bachelard has called "phenomenotechnique." But this allows us only to describe the equipment of the group at one point in time. At some earlier point, each item of equipment had been a contentious set of arguments in a neighbouring discipline. Consequently, one cannot take for granted the difference between "material" equipment and "intellectual" components of laboratory activity: the same set of intellectual components can be shown to become incorporated as a piece of furniture a few years later. In the same way, the long and controversial construction of TRF was eventually superceded by the appearance of TRF as a noncontroversial material component in other assays. Similarly, we briefly indicated, at the end of Chapter 5, how investments made within the laboratory were eventually realised in clinical studies and in drug industries. In order to emphasise the importance of the time dimension, we shall refer to the above process as *materialisation,* or *reification* (Sartre,1943). Once a statement stabilises in the agonistic field, it is reified and becomes part of the tacit skills or material equipment of another laboratory.[6] We shall return later to this point.

The fourth concept upon which we have drawn is that of *credibility* (Bourdieu, 1976). We used credibility to define the various investments made by scientists and the conversions between different aspects of the laboratory. Credibility facilitates the synthesis of economic notions (such as money, budget, and payoff) and epistemological notions (such as certitude, doubt, and proof). Moreover, it emphasises that information is *costly.* The cost-benefit analysis applies to the type of inscription devices to be employed, the career of scientists concerned, the decisions taken by funding agencies, as well as to the nature of the data, the form of paper, the type of journal, and to readers' possible objections. The cost itself varies according to the previous investments in terms of money, time, and energy already made.[7] The notion of credibility permits the linking of a string of concepts, such as accreditation, credentials and credit to beliefs ("credo," "credible") and to accounts ("being accountable," "counts," and "credit accounts"). This provides the observer with an homogeneous view of fact construction and blurs arbitrary divisions between economic, epistemological, and psychological factors.[8]

The fifth concept used in our argument, albeit somewhat programmatically, is that of *circumstances* (Serres, 1977). Circumstances (that which stands around) have generally been considered irrelevant to the practice of science.[9] Our argument could be summarised as an attempt to demonstrate their relevance. Our claim is not just that TRF is surrounded, influenced by, in part depends on, or is also caused by circumstances; rather, we argue that science is entirely fabricated out of circumstance; moreover, it is precisely through specific localised practices that science appears to escape all circumstances. Although this has already been demonstrated by some sociologists (for example, Collins, 1974; Knorr, 1978; Woolgar, 1976), the concept of circumstances has also been developed from a philosophical perspective by Serres (1977). Chapter 2 is an analysis of the circumstances which make stable objects possible in neuroendocrinology; Chapter 3 shows in which networks TRF is able to circulate outside the laboratory in which it was originally constructed; at the end of Chapter 4 we record how the same holds true for the extension of somatostatin. We also point out in Chapter 4 how daily conversations constantly feature local or idiosyncratic circumstances. Finally, in Chapter 5, we use the notion of positions in order to account for the circumstancial character of careers. Rather than being a structure or an ordered pattern, a field consists only of positions which influence each other in a way which is not itself orderly (see pp. 211ff). The notion of position enables us to talk about the "right" time, or the "right" assay, or in Habermas's (1971) terms, to replace the historicity in science (Knorr, 1978).

The sixth and final concept upon which we have drawn is *noise* (or, more exactly, the ratio of signal to noise), which is borrowed from information theory (Brillouin, 1962). Its application to an understanding of scientific activity is not new (Brillouin, 1964; Singh, 1966; Atlan, 1972), but our usage is very metaphorical. We have not, for example, attempted to calculate the signal to noise ratio produced by the laboratory. But we have retained the central idea that information is measured against a background of equally probable events, or as Singh (1966) puts it:

> We measure the information content of a message in any given ensemble of messages by the logarithm of the probability of its occurrence. This way of defining information has an earlier precedent in statistical mechanics where the measure of entropy is identical in form with that of information (Singh, 1966: 73).

The concept of noise fits closely with our observations of participants busily reading the written tracts of inscription devices (see Chapter 2, pp. 48ff). The notion of equally probable alternatives also allowed us to describe the final construction of TRF in Chapter 3: the import of mass spectrometry delimited the number of probable statements. In Chapter 5, the notion of demand, which allowed us to develop the idea of a market for information and to permit the operation of the credibility cycle, was based on the premise that any decrease in the noise of one participant's operation enhances the ability of another participant to decrease noise elsewhere.

The result of the *construction* of a fact is that it appears unconstructed by anyone; the result of rhetorical *persuasion* in the agonistic field is that participants are convinced that they have not been convinced; the result of *materialisation* is that people can swear that material considerations are only minor components of the "thought process"; the result of the investments of credibility, is that participants can claim that economics and beliefs are in no way related to the solidity of science; as to the *circumstances*, they simply vanish from accounts, being better left to political analysis than to an appreciation of the hard and solid world of facts! Although it is unclear whether this kind of inversion is peculiar to science,[10] it is so important that we have devoted much of our argument to specifying and describing the very moment at which inversion occurs.

Having summarised the main arguments of the preceding chapters, it is important now to show how they are related because the concepts above have been borrowed from several different fields.

Let us start with the concept of noise. For Brillouin, information is a relation of probability; the more a statement differs from what is expected, the more information it contains. It follows that a central question for any participant advocating a statement in the agonistic field is how many alternative statements are equally probable. If a large number can easily be thought of, the original statement will be taken as meaningless and hardly distinguishable from others. If the others seem much less likely than the original statement, the latter will stand out and be taken as a meaningful contribution.[11] When a laboratory member reads a peak on an amino acid analyser, for example (Photograph 9), he first needs to ascertain whether or not he can convince himself (or others)[12] that the peak is different from the background noise. As we have seen, this depends in part on his colleagues. If his claim, "look at this peak," meets with the response, "there is no peak, it is simply noise, you might just as well say that the peak is this little blurr at the other side" (see Photograph 8), his statement has no informative value (in this context).

The sentence which threatens to dissolve all statements (and careers) takes the conditional form: "*but you might as well say* that it is . . . " and precedes a list of equally probable statements. The outcome of this formulation is often the dissolution of the statement in noise. So the objective of the game is to carry out all possible manoeuvres which might force the scientist (or colleagues) to admit that alternative statements are not equally plausible. We discussed some of the manoeuvres in Chapters 3 and 4. One common manoeuvre is that of *construction*. By showing colleagues, two, rather than one, peaks of an amino acid analysis, or by increasing the distance between the peak and base line, the difference between the various possible statements will also be increased. By being sufficiently convincing, people will *stop* raising objections altogether, and the statement will move toward a fact-like status. Instead of being a figment of one's imagination (subjective), it will become a "real objective thing," the existence of which is beyond doubt.[13]

The operation of information construction, then, transforms any set of equally probable statements into a set of *unequally* probable statements. At the same time, this operation draws upon the activities of persuasion (agonistic) and of writing (construction) in order to increase the signal to noise ratio.

How can inequality be introduced into a set of equally probable statements in such a way that a statement is taken to be more probable than all the alternatives? The technique most frequently used by our scientists was that of *increasing the cost* for others to raise equally probable alternatives. In Chapter 3, for example, we showed that the imposition of new standards on the field of releasing factors effectively ruined competitors' efforts. Similarly, when Burgus used mass spectrometry to make a point, he made it difficult to raise alternative possibilities because to do so would be to contest the whole of physics. Once a slide has been shown with all the lines of the spectrum corresponding to one atom of the amino acid sequence, no one is likely to stand up and object.[14] The controversy is settled. But if a slide is presented which shows the spots of a thin-layer chromatography, ten chemists will stand up and assert that "this is not a proof." The difference, in the second case, is that any chemist can easily find fault in the method used (but see the Donohue episode, p. 171).

This point would clearly be tautological but for the central notion of materialisation or *reification* which we defined earlier and can now use at its best. The mass spectrometer is the reified part of a whole field of physics; it is an actual piece of furniture which incorporates the majority of an earlier body of scientific activity. The cost of disputing the generated results of this inscription device has been enormous. Indeed, this explains by Guillemin and Burgus strived from the beginning to "get at the mass spectrometer." In the case of thin layer chromatography, however, very little earlier interpretative work has been reified. Consequently, it is easy to contest any step in arguments based on a chromatograph and to propose an alternative argument. Once a large number of earlier arguments have become incorporated into a black box,[15] the cost of raising alternatives to them becomes prohibitive. It is unlikely, for example, that anyone will contest the wiring of the computer shown in Photograph 11, or the statistics on which the "t" test is based, or the name of the vessels in the pituitary.

The operation of black-boxing is made possible by the availability of credibility (Ch. 5). As we argued earlier, credibility is a part of the wider phenomenon of credit, which refers to money, authority, confidence and, also marginally, to reward. The first question raised when a statement is proposed, is how much the statement and/or its author can be credited. This question is directly analogous to the question of cost mentioned above: what sort of investments should be made so as to fabricate a statement of equal probability to that of a competitor? In a million-dollar business like the sequencing of TRF, the chances are that no alternative statement is feasible. The constraints are such that no investment could possibly match those already made. Consequently, statements which are already credited will be taken for granted. In addition, they will be used to make points in other laboratories. This is the nature of the market defined in Chapter 5. No matter whether this taken-for-granted peptidic structure takes the form of a nonproblematic argument or of a white powder sample, the only important question is whether borrowing it (or buying it) will make it more difficult for a competitor to contest statements.

Of course, the concepts of cost, reification, and credit have to be understood in the light of our earlier argument: everything which has been accepted, *no matter for what reason*, will be reified so as to increase the cost of raising objections. For instance, the standing of one scientist might be such that when he defines a problem as important, no one feels able to object that it is a trivial question; consequently, the field may be moulded around this important question, and funds will be readily forthcoming. In the Donohue episode, chemists' preference for the enol form for the four DNA bases was stabilised and reified in textbooks, such that it was more difficult for Watson to doubt it or simply to object that the keto form was equally probable. The cost-benefit analysis will vary according to the prevailing *circumstances*, so no general rules can be established. The style of an article can make it more difficult for the reader not to believe in it; the qualification of statements can disarm readers' objections; for another audience, documentation through the use of footnotes can add conviction; competitors can even be silenced by imprisonment or fraud (Lecourt, 1976). The major rule of the game is to assess the cost of investments compared with their likely return; the game is not played according to a set of ethical rules, which a superficial examination reveals.[16]

The portrayal resulting from the above combination of concepts used throughout our argument has one central feature: the set of statements considered too costly to modify constitute what is referred to as reality. Scientific activity is not "about nature," it is a fierce fight to *construct* reality. The *laboratory* is the workplace and the set of productive forces, which makes construction possible. Every time a statement stabilises, it is reintroduced into the laboratory (in the guise of a machine, inscription device, skill, routine, prejudice, deduction, programme, and so on), and it is used to increase the difference between statements. The cost of challenging the reified statement is impossibly high. Reality is secreted.[17]

So far we have summarized the main points of our argument by showing how six of the major concepts we have used are related and, finally, by zooming in on the notion of laboratory from which we started in the second chapter. There is, however, an alternative way of describing laboratory life which draws primarily on one single concept.

Order From Disorder

The transformation of a set of equally probable statements into a set of unequally probable statements amounts to the creation of order (Brillouin, 1962; Costa de Beauregard, 1963; Atlan, 1972). Let us now provide a new account of laboratory life using the notion of order together with Brillouin's famous mythical character: Maxwell's demon. The simplest version is the following (Singh, 1966):

A demon placed in a cold oven would be able to increase the amount of heat by allowing the swifter molecules to gather in one part of the oven and by keeping them there. In order to do this, the demon needs information about the state of the molecules, a small trap which will let them come or go depending on their quality, and an enclosure in which to prevent the sorted molecules from escaping and returning to their random state. We now know that the demon himself consumes a small amount of energy in doing his work. "It is impossible to get something for nothing, even information," as the saying goes.

This account provides an illuminating analogy with what goes on in the laboratory. We have already seen the laboratory to be an enclosure where previous work is gathered. What would happen if this enclosure was opened? Imagine that the following experiment was carried out by our observer. Entering the deserted laboratory at night, he opens one of the large refrigerators shown in Photograph 2. As we know, each sample on the racks corresponds to one stage of the purification process and is labelled with a long code number which refers back to the protocol books. Taking each sample in turn, the observer peels off the labels, throws them away and returns the naked samples to the refrigerator. Next morning, he would doubtless witness scenes of extreme confusion. No one would be able to tell which sample was which. It would take up to five, ten, and even fifteen years (the time it took to label the samples) to replace the labels—unless, of course, chemistry techniques had advanced in the mean time. As we stated earlier, any sample might equally well be any other. In other words, the disorder, or more precisely the entropy, of the laboratory would have increased: anything could be said about each and every sample. This nightmarish experiment highlights the importance of the trapping system for any competent Maxwell's demon wishing to decrease disorder.[18]

At this point, we can perhaps do justice to the apparently strange notion of *inscription* introduced in Chapter 2. Our argument there was that writing was not so much a method of transferring information as a material operation of creating order. Let us illustrate the importance of writing by reference to an experiment undertaken by the observer during his stay in the laboratory. As we mentioned in Chapter 1, the sociologist worked as a technician during his participant observation. Fortunately for us, the observer turned out to be an extremely bad technician in a very efficient laboratory. Consequently, his deficiencies highlighted the roots of his informants' competence. One of the most difficult tasks was the dilution and addition of doses to the beakers. He had to remember in which beaker he had to put the doses, and made a note, for example, that he had to put dose 4 in beaker 12. But he found that he had forgotten to make a note of the time interval. With pipette half lifted, he found himself wondering whether he had *already* put dose 4 in beaker 12. He blushed, trying to remember whether he had made a note before or after the actual action took place; obviously, he had not made a note of when he had made a note! He panicked and pushed the piston of the pasteur pipette into beaker 12. But maybe he had now put *twice* the dose into the beaker. If so, the reading would be wrong. He crossed out the figure. The observer's lack of training meant that he continued in this fashion. Not surprisingly, the resulting points exhibited wide scatter. A day's work had been lost. It is necessary to be a technician, and an incompetent one at that, in order fully to appreciate the practical miracle (in Boltzmann's sense of the word) which gives rise to a standard curve. A wealth of invisible skills underpin material inscription. Every curve is surrounded by a flow of disorder, and is only saved from dissolution because everything is written or routinised in such a way that a point cannot as well be in any place of the log paper. But the unhappy observer was not party to these constraints! Instead of creating more order, he had only succeeded in creating less; and, in the meantime, he had used up animals, chemicals, time, and money.

Even insecure bureaucrats and compulsive novelists are less obsessed by inscriptions than scientists. Between scientists and chaos, there is nothing but a wall of archives, labels, protocol books, figures, and papers.[19] But this mass of documents provides the only means of creating more order and thus, like Maxwell's demon, of increasing the amount of information in one place. So it is easy to appreciate their obsession. Keeping track is the only way of seeing a pattern emerge out of disorder (Watanaba, 1969). It might be impossible to differentiate any of a thousand equally active peptides out of a soup of unpurified brain extracts. If assays designed to separate out one of these peptides were carefully carried out but not recorded, the technicians would have to start all over again; there would be no way of discriminating between statements because there would be no superimposition of traces and consequently no construction of an object. When, by contrast, a series of curves have been recorded, and it is possible to spread them out on the large library table and ponder them, then an object is in the process of construction. Objects appear because of the constant process of sorting. Thin readable traces (produced by the inscription devices) are recorded and this creates a pocket of order in which not everything is equally probable. In view of eight years' worth of documents and a million dollars' worth of equipment, the range of possible statements which can be made about the structure of TRF is restricted. The cost of selecting a statement from outside this range is prohibitive.

Maxwell's demon provides a useful metaphor for laboratory activity because it shows both that order is *created* and that this order in no way preexists the demon's manipulations. Scientific reality is a pocket of order, created out of disorder by seizing on any signal which fits what has already been enclosed and by enclosing it, albeit *at a cost*. In order fully to explore the force of this model, however, it is necessary to examine the relation between order and disorder in more detail. Disorder is not only the noise in which statements made by inefficient technicians are dissolved; paradoxically, the laboratory is also involved in the production of disorder. By recording all events and keeping traces from all the inscription devices, the laboratory overflows with computer listings, data sheets, protocol books, diagrams, and so on. Even if it successfully resists the outside disorder, the laboratory itself generates disorder within its enclosure. The noise of thousands of brain extracts, is replaced by the noise of accumulated data. Information again seems like the elusive needle in a haystack. No patterns emerge. Participants' solution to this danger is selectively to eliminate material from the mass of accumulated data. Here is the importance of the statements, the genealogy of which we outlined in Chapter 2. The problem is not now to discern a peak from the background noise (the baseline), but to read a sentence out of the mass of gathered peaks and curves. One particular curve is selected, cleaned up, put on a slide and shown around in conjunction with the statement: "Stress simultaneously releases ACTH and Beta Endorphine." This statement stands out of and for the mass of figures. A paper begins to be drafted, which constitutes a second-degree enclosure (an enclosure represented in Fig. 2.1 by the laboratory partitions).

Sorting, picking up and enclosing are costly operations, and they are rarely successful; any slackening can once again drown a statement in confusion. This is more so because a statement exists, not by itself, but in the agonistic field (or market, Ch. 5) made up of the laboratories striving to decrease their own noise. Is the statement going to stand out in the field or will it merely once again be drowned in the mass of literature on the subject? Perhaps it is already redundant, or simply wrong. Perhaps it will never be picked out from the noise. The laboratory's production process again seems chaotic: statements have to be pushed, forced into the light, defended against attack, oblivion, and neglect. Very few statements are seized upon by everyone in the field because their use entails an enormous economy in the manipulation of data or statements (Brillouin, 1962: Ch. 4). These statements are said to "make sense" or "to explain a lot of things" or to allow a dramatic decrease in the noise of one inscription device: "now we can obtain reliable data." Such very rare events, the sorting of facts from the background noise, are often heralded by the Nobel Prizes and a flourish of trumpets.

Maxwell's demon *creates* order. This analogy not only provides a way of summarising and relating the main concepts used in our earlier description of laboratory activity; it also helps answer the objection that we have not explained why a controversy becomes resolved, or why a statement stabilises. But this objection only has meaning in so far as it is assumed that order somehow preexists its "revelation" by science, or in some way results from something other than disorder. This basic philosophical assumption has recently been challenged, and our intention in the next part of this chapter is to show what light is shed on laboratory activity if such an assumption is modified. To do this in full would entail going beyond the usual range of argument in sociology of science, and certainly beyond the scope of this monograph. We therefore restrict our discussion to one further analogical description of the laboratory.

Figure 6.1 shows three stages of a game of "go" as related by Kawabata (1972). The game of go starts from an empty board to which

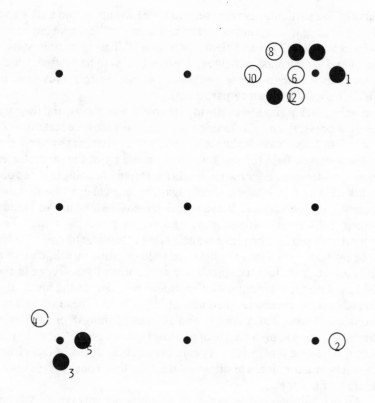

Figure 6.1a

stones are added in successive moves. The added stones do not move around the board as, for example, in chess. Consequently, the first moves are almost entirely contingent (Figure 6.1a). As the game progresses, however, it becomes less and less easy to play anywhere; as in the agonistic field, the results of earlier play transforms the set of future possible moves. Not all moves are equally possible (Figure 6.1b). Indeed, some are totally impossible (for example, white cannot play on the upper left hand corner), others are less likely, and some are almost necessary (for example, play at 64 after 63 in Fig. 6.1c). As in the agonistic field, the changing pattern is not orderly: in the lower right hand corner or in the middle of the board, it is possible to play almost anywhere; but the situation in the left hand corner is definitively settled. A territory may or may not be defended according to the

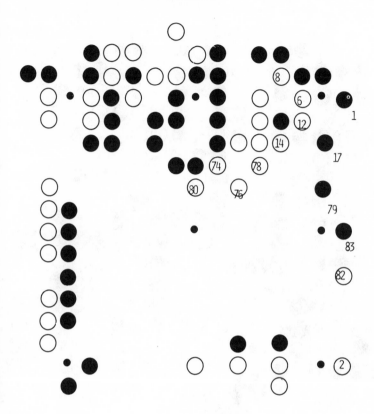

Figure 6.1b

pressures exerted by the opponent. The game ends when all territory has been appropriated (Figure 6.1c) and all disputed territories have been settled (for example, the stones at the top). From an entirely contingent beginning, the players arrive (without the use of external or preexisting order) at a final point in the game where certain moves are *necessary*. In principle, any individual move could be made anywhere; in practice, the cost of spurning what appears the necessary move is prohibitive.[20]

The relationship between order and disorder, which underpins our account of the construction of facts, is very familiar to biologists (Orgel, 1973; Monod, 1970; Jacob, 1977; Atlan, 1972). That life is an orderly pattern emerging from disorder through the sorting of random mutations, is the stock in trade of the biological representation of life.

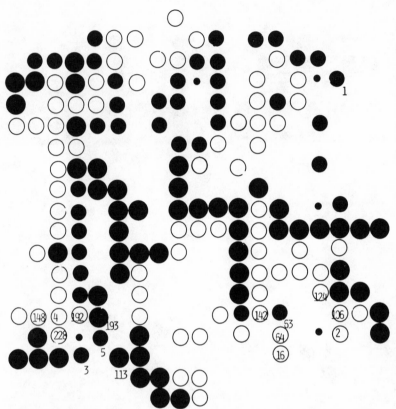

Figure 6.1c
Figure 6.1a-c is taken from the novel by Kawabata (1972). It shows three moments in the unfolding of a game of "Go." 6.1a shows the board at the 10th move; 6.1b at the 80th; and 6.1c at the end. The game of "Go" provides a model for the construction of orderly but unpredictable forms. The same stones appear in each of the three diagrams. The most important moves are signalled by numbers.

For Monod, for example, chance (disorder) and necessity (a sorting mechanism) are sufficient to account for the emergence of complex organisation. Reality is constructed out of disorder, without the use of any preexisting representation of life. Many of the laboratory members themselves used terms such as chance, mutation, niches, disorder, and tinkering (Jacob, 1977) to account for life itself. But sociologists of science seem extremely reluctant to introduce similar concepts to account for the construction of reality.[21] After all, the construction of reality might be no more complex than the generation of organisms.

The three brief analogies drawn above (Maxwell's devil, the game of Go, and Monod's notion of chance and necessity) were intended simply as a way of familiarising the reader with the slight modification of background which is well known in many other disciplines, but which seems to have escaped the attention of analysts of science.

It is part of our world view that things are ordered, that order is the rule, and that disorder should be eliminated wherever possible. Disorder always has to be eliminated from politics and ethics as well as from science. It is also part of our world view that only from disorder can an orderly pattern emerge. These assumptions have recently been challenged by several philosophers, especially Michel Serres, who, in turn, have been greatly influenced by authors such as Brillouin and Boltzmann and by new developments in biology. Their argument is that these assumptions be inverted, that disorder be considered the rule and order the exception. This argument has been familiar since life was first considered to be a neguentropic event which fed off the much larger and opposite trend towards entropy. Recently, this picture has been extended to include science itself as a marginal case of a certain type of social organism, a particular but not a peculiar case of neguentropy (Monod, 1970; Jacob, 1977; Serres, 1977a; 1977b). For our purposes, the interesting part of this argument is the contention that the construction of order relies upon the existence of disorder (Atlan, 1972; Morin, 1977). If one accepts this suggested modification, it is possible to discern a marked convergence between our approach and apparently disparate approaches to the social study of science.[22] Let us consider four such approaches.

Firstly, the history of science can be characterised as demonstrating the chain of circumstances and unexpected events leading to this or that discovery. However, this mass of events is not easily reconciled with the solidity of the final achievements. This is one reason why the context of justification is so frequently opposed to the context of discovery. With the above modification of our background assumption, this opposition is no longer necessary (Feyerabend, 1975; Knorr, 1978). To use Toulmin or Jacob's analogies, if life itself results from tinkering and chance, it is surely not necessary to imagine that we need more complex principles to account for science. The "*événemential-isation*" (Foucault, 1978) of science made by historians penetrates the core of fact construction. Secondly, sociologists have demonstrated the importance of informal communication in scientific activity. This well-documented phenomenon takes on a new meaning against the newly modified assumption: the production of new information is necessarily obtained by way of unexpected meetings, through old boy networks and by social proximity. The informal flow of information does not contradict the orderly pattern of formal communication. Instead, as we have suggested, much informal communication derives its structure from its constant referral to the substance of formal communication. Nonetheless, informal communication *is the rule*. Formal communication is the exception, as an *a posteriori* rationalisation of the real process. Thirdly, citation analysts have demonstrated the extensive waste of energy in scientific activity. Most published papers are never read, the few that are read are worth little, and the remaining 1 or 2 percent are transformed and misrepresented by those who use them. But this waste no longer appears paradoxical if we accept the hypothesis that order is an exception and disorder the rule. Few facts emerge from the substantial background noise. The circumstances of discovery and the process of informal exchange are both crucial to the productive process: they are what allows science to exist at all. Finally, growing sociological interest in the details of negotiation between scientists has revealed the unreliability of scientists' memories and the inconsistency of their accounts. Each scientist strives to get by amid a wealth of chaotic events. Every time he sets up an inscription device, he is aware of a massive background of noise and a multitude of parameters beyond his control; every time he reads *Science* or *Nature*, he is confronted by a volume of contradictory concepts, trivia, and errors; every time he participates in some controversy, he finds himself immersed in a storm of political passions. This background is everpresent, and it is only rarely that a pocket of stability emerges from it. The revelation of the diversity of accounts and inconsistency of scientific arguments should therefore come as no surprise: on the contrary, the emergency of an accepted fact is the rare event which should surprise us.

A New Fiction For Old?

We have so far in this chapter summarised the arguments of the former chapters, showed how they are related through the notion of the construction of order out of disorder and linked them to what has been done in sociology of science. We shall now summarise the methodological problems encountered in the course of our argument, looking in particular at the thorny issue of the status of our own account. What is the basis for our claim that scientists produce order from disorder? Obviously, our own account cannot escape the conditions of its own construction. From what kind of disorder does our account emerge? In which agonistic field are we to put together differences between fiction and fact?

Throughout the argument, we have stressed the importance of avoiding certain distinctions commonly adopted by analysts of scientific activity. In Chapter 1, we refused to accept the distinction between social and technical issues; in Chapter 2, we had to suspend any given distinction of nature between facts and artefacts; in Chapter 3, we demonstrated that the difference between internal and external factors was a consequence of the elaboration of facts rather than a given starting point for understanding their genesis; in Chapter 4, we argued for the suspension of a priori distinctions between common sense and scientific reasoning; even the distinction between "thought" and craftwork needed to be avoided as an explanatory resource because it appeared to be the *consequence* of scientific work in the laboratory; similarly, in Chapter 5, we argued that the notion of scientists as individuals was the consequence of the appropriation conflicts within the laboratory.

Stylistically, the replacement and avoidance of these obsolete distinctions presented severe difficulties. In allying our discussion to each of certain literary genres (for example, the "historical" discussion of Ch. 3), we found ourselves constrained by using terminology which tended to reintroduce these distinctions. For this reason, it was necessary to look carefully at our own usage of words. For example, the term social has connotations which make it difficult to avoid importing distinctions, such as that between social and technical. Similarly, the term familiar obscures the particular sense with which we wanted to apply the notion of an anthropology of science. In Chapter 3, in particular, we had to resist terminology commonly employed in historical accounts because it had the tendency of transforming constructed facts into "discovered" facts. In Chapter 4, the use of the expression "I had an idea" or the tautological use of "scientific" was sufficient to destroy the tenor of our argument. Consequently, it was necessary to dispute some of the terms used by epistemologists. By employing the word credit and by exploring its various different meanings, we circumvented some of the distinctions which usually come to mind when one uses terms such as strategy, motivations, and careers.

We have thus tried to exercise some care in discriminating between the kinds of terms and distinctions which might jeopardise our account of laboratory life. However, we have as yet to clarify what differentiates our account of laboratory life from those routinely produced by scientists. Is there any essential distinction between the nature of our own construction and that used by our subjects? Emphatically, the answer must be no. Only by rejecting the possibility of this last distinction can the arguments of this chapter cohere. The notion of creating order from disorder applies as much to the construction of our own account as to that of the laboratory scientists. How then do we know how they know?

How have we built up our account of fact production whereby laboratory scientists get by with fictions which they push as hard as they can in the agonistic field?

If we return to the situation (described in Ch. 2) where the naive observer visited the "strange" laboratory, it is clear that he constructed his preliminary accounts out of disorder. He neither knew what to observe, nor the names of the objects in front of him. In contrast to his informants, who exhibited confidence in all their actions, our observer felt distinctly uneasy. He found himself wondering where to sit, when to stand, how to present himself, and what questions to ask. A flood of gossip, anecdotes, lectures, explanations, impressions, and feelings emerged from his initial daily contact with the laboratory. Subsequently, however, he began to set up a crude inscription device to monitor these data. He found himself as observer connected up to a screen (his notebooks), the effects being recorded by means of amplification (such as his definition of assays). But these first "socioassays" were extremely noisy and chaotic. The early notebooks reveal the confusion of the first recordings: trivia, generalities, noise, and more noise.

The observer was obliged to create some stable pockets of order out of this flood of impressions. He attempted this, first by a crude imitation of the method of his informants: he plotted time on one axis of a piece of graph paper and wrote the names of the scientists on the other. Armed with a watch, he inscribed who did what and when. In this way he began to produce ordered information. In another instance, he distilled the pattern of citations received by group members from the mass of citation data in the SCI. Like any conscientious Maxwell's devil, he filtered the names he required, counted the citations and inscribed them in columns. One result was Figure 5.3: a relatively modest achievement, admittedly, but one which granted him a brief moment of contentment. On the basis of this result, he could make a statement: when his informants objected that the claim was nonsensical, he was then able to produce the figure and this had the effect of quietening his audience, at least temporarily.

In the course of a few months, our observer accumulated a sizeable body of similar figures, documents, and other notes. In terms of the analogy with "go" he began to fill his board with random moves. Consequently, as he progressed further, he realised that it was no longer possible to make just *any* statement on the basis of this accumulated material. In addition, our observer found himself able either to counter or support some of the arguments in the science studies literature. He could also transform them into artefacts or facts with the use of the objects he had begun to amass. He began to write articles and to operate in his own agonistic field. At this stage, however, his accounts were so weak that any other account seemed equally plausible. Moreover, his informants flooded him with contradictory examples and argued for alternative interpretations.

By returning to the initial stages of the study, then, we can discern an essential similarity between the methods of the observer and his informants. Even so, it is not clear who was imitating whom. Were the scientists imitating the observer, or vice versa?

As mentioned earlier, part of the observer's experience involved his participation as laboratory technician. From time to time he could don a white coat, go into the biassay room, and set up an assay for the melanotropin stimulating hormone (MSH) instead of drawing citation curves and transcribing interviews. (MSH darkens frog skin, as measured by variations of light in a reflectometer.)

The observer had his protocol book and an empty data sheet in front of him. He seized the jumping frogs, beheaded, and flayed them, and finally immersed thin sections of skin in the beakers. He placed each of the beakers over a source of light and took readings from the reflectometer, which he then wrote down. By the end of the day, he had accumulated a small stack of figures which could be fed into the computer (Photograph 11). After this he was left only with standard deviations, levels of significance, and means in the computer listing. On the basis of these he drew a curve and, taking it into his boss's office, argued about the slight differences or similarities in the curve in order to make a point.

Some similarities between the construction of the citation curve and that of the standard curve for MSH are obvious. Thus, the following features are common to both activities. Inscription devices were set up; five or ten names were singled out of the millions in the SCI (only a few pieces of skin were taken from the complexity of the frog organism); the investigator placed a premium on those effects which were recordable; the data were cleaned up so as to produce peaks which were clearly discernible from the background; and, finally, the resulting figures were used as sources of persuasion in an argument. These similarities make it difficult to maintain that there is any fundamental difference between the methods of "hard" and "soft" science.

The similarity of his two roles began to prove unnerving. Our observer sometimes felt himself completely assimilated into "his" laboratory: he was addressed as "doctor," possessed protocol books and slides, submitted papers, met colleagues at congresses and busied himself setting up new inscription devices and filling in questionnaires. On the other hand, he was painfully aware of the enormous distance between the apparent solidity of his informants' constructions and his own. In order to study half a gram of brain extract, they had at their disposal tons of material, millions of dollars, and a large group of some forty people; in order to study the laboratory, our observer was alone. At the bench, working on the MSH assay, people would constantly peer over his shoulder and criticise him ("don't hold your pipette like that"; "let me redo your dilution"; "check this reading again") or direct his attention to one of the sixty articles written about the assay.[23] While tinkering a few makeshift methods for analysing the work of the laboratory, he had few general contacts and no precedent upon which he felt he could build. The scientists had a laboratory, in which were gathered all the stable objects of their field, and free access to the object under construction; the observer had no such resources. Moreover, he had to settle in the laboratory used as a resource by the scientists and to beg information as a stranger, a foreigner, and a layman.

The difference in credibility accorded the observer's and the informants' constructions corresponds directly to the extent of prior investments. Occasionally, when members of the laboratory derided the relative weakness and fragility of the observer's data, the observer pointed out the extent of the imbalance between the resources which the two parties enjoyed. "In order to redress this imbalance, we would require about a hundred observers of this one setting, each with the same power over their subjects as you have over your animals. In other words, we should have TV monitoring in each office; we should be able to bug the phones and the desks; we should have complete freedom to take EEGs; and we would reserve the right to chop off participants' heads when internal examination was necessary. With this kind of freedom, we could produce hard data." Inevitably, these kinds of remarks sent participants scurrying off to their assay rooms, muttering darkly about the "Big Brother" in their midst.

Gradually, the observer gained confidence in his work: he was both adding to the stockpile of inscriptions in his office and beginning to realise that there was nothing special or mysterious about the difference between his activity and that of his informants. The essential similarity was that both were engaged in craftwork; differences could be explained in terms of resources and investments, and without recourse to exotic qualities of the nature of the activity. Consequently, the observer began to feel less intimidated. When his informants were interpreting traces on the library table, for example, they really seemed little different to him; they pondered diagrams, putting some aside, evaluating the strength of others, seizing on weak analogical links, and so slowly constructed an *account*. At the same time, the observer was writing a fictional account on the basis of makeshift curves and documents. Informants and observer shared participation in the art of interpreting confused texts (texts comprising slides, diagrams, other paper, and curves) and of writing persuasive accounts.[4]

Our account of fact construction in a biology laboratory is neither *superior nor inferior* to those produced by scientists themselves. It is not superior because we do not claim to have any better access to "reality," and we do not claim to be able to escape from our description of scientific activity: the construction of order out of disorder at a cost, and without recourse to any preexisting order. In a fundamental sense, our own account is no more than *fiction*.[25] But this does not make it inferior to the activity of laboratory members: they too were busy constructing accounts to be launched in the agonistic field, and loaded with various sources of credibility in such a way that once convinced, others would incorporate them as givens, or as matters of fact, in their own construction of reality. Nor is there any difference in the sources of credibility upon which they and we can draw so as to force people to drop modalities from proposed statements. The only difference is that *they have a laboratory*. We, on the other hand, have a text, this present text. By building up an account, inventing characters (for example, the observer of Ch. 2), staging concepts, invoking sources, linking to arguments in the field of sociology, and footnoting, we have attempted to decrease sources of disorder and to make some statements more likely than others, thereby creating a pocket of order. Yet this account itself will now become part of a field of contention. How much further research, investment, redefinition of the field, and transformation of what counts as an acceptable argument are necessary to make this account more plausible than its alternatives?

NOTES

1. This point has been made frequently by Bachelard (for example, 1934; 1953). However, his interest in demonstrating the "mediations" in scientific work was never extended. His "rational materialism," as he put it, was more often than not the basis for distinguishing between science and "prescientific" ideas. His exclusive interest in "la coupure épistémologique" prevented him from undertaking sociological investigations of science, even though many of his remarks about science make better sense when set within a sociological framework.

2. From the outset, the observer was struck by the almost absurd contrast between the mass of the apparatus and the minute quantities of processed brain extract. The interaction between scientific "minds" and "nature" could not adequately account for this contrast.

3. In a different context, the importance of the stakes may vary. For example, the importance of somatostatin for the treatment of diabetes ensures that each of the group's articles is carefully checked. In the case of endorphine, by contrast, any article (no matter what the wildness of its conjectures) will initially be accepted as fact.

4. On his first day in the laboratory, the observer was greeted with a maxim which was constantly repeated to him in one or another modified form throughout his time in the field: "The truth of the matter is that 99.9% (90%) of the literature is meaningless (crap)."

5. We base this argument on several conversational exchanges which took place between lawyers and scientists. Unfortunately, we are not able to make explicit use of this material here.

6. It is crucial to our argument that anything can be reified, no matter how mythical, absurd, whimsical, or logical it might seem either before or after the event. Callon (1978), for example, has shown how technical apparatus can incorporate the outcome of totally absurd decisions. Once reified, however, these decisions take the role of premise in subsequent logical arguments. In more philosophical terms, one cannot understand science by accepting the Hegelian argument that "real is rational."

7. But for a few pages in Lacan (1966) and some indirect hints by Young (n.d), a psychoanalytic understanding of these kinds of energy investments is as yet undeveloped.

8. For example, Machlup (1962) and Rescher (1978) have attempted to understand the information market in economic terms. However, their approach extends rather than transforms the central notion of economic investment. By contrast, Bourdieu (1976) and Foucault (1978) have outlined a general framework for a political

economy of truth (or of credit) which subsumes monetary economics as one particular form of investment.

9. The philosophical enterprise can be characterised as an attempt to eliminate any trace of circumstances. Thus, the task of Socrates in Plato's *Apology of Socrates* is to eliminate circumstances included in the definition of activity provided by the artist, the lawyer, and so on. Such elimination is the price which has to be paid in order to establish the existence of an "idea." Sohn Rethel (1975) has argued that such philosophical operations were essential for the development of science and economics. It could be argued, therefore, that the task of reconstructing circumstances is fundamentally hampered by the legacies of a philosophical tradition.

10. Barthes argues that this kind of transformation is typical of modern economics. It is thus possible that there is some useful similarity between Marx's (1867) notion of fetishism and the notion of scientific facts. (Both fact and fetish share a common etymological origin.) In both cases, a complex variety of processes come into play whereby participants forget that what is "out there" is the product of their own "alienated" work.

11. Brillouin uses the word likely in a counterintuitive way. It is only if a statement is unlikely that it contains information since its distance from the background of equally probable statements is very great. In ordinary language, however, we might say that people believe a statement when it is more likely than the others. The reason for this apparent contradiction is that information is nothing but a ratio of signal to noise.

12. In the course of our discussion, we have tried to minimise distinctions between convincing ourselves and convincing others. In interviews the continuous shortcuts between the two were so common ("I wanted to be sure, and I did not want W to stand up and contradict me"), that we gave up making this artificial distinction. Our experience suggests that, perhaps in the most secret part of his consciousness, a scientist argues with the whole agonistic field and anticipates every single one of his colleagues' potential objections.

13. This formulation closely matches scientists' own impression of a messy field: it is a field in which you can say *anything* or, more precisely, in which *anyone can* equally well say anything.

14. This is not to say that it is impossible in principle to contest the argument based on the use of a mass spectrometer. But the cost of modifying the basis of the theory is so high that, in practice, no one will challenge it. (The exception, perhaps is in the case of a scientific revolution.) The difference between what is possible in principle and what can be done in practice is the lynchpin of our argument. As Leibnitz put it: "Everything is possible, but not everything is compossible." The process by which the realm of compossibility is extended was explored in Chapter 3. The mass spectrometer is no more truthful than thin-layer chromotography; it is simply more powerful.

15. The term "black box" also brings to mind Whitley's (1972) argument that sociologists of science should not treat the cognitive culture of scientists as a self-contained entity immune from sociological investigation. Although we sympathise with this view, Whitley misses a crucial point. The activity of creating black boxes, of rendering items of knowledge distinct from the circumstances of their creation, is precisely what occupies scientists the majority of the time. The way in which black boxing is done in science is thus an important focus for sociological investigation. Once an item of apparatus or a set of gestures is established in the laboratory, it becomes very difficult to effect the retransformation into a sociological object. The cost of revealing

sociological factors (the cost, for example, of portraying the genesis of TRF) is a reflection of the importance of the black boxing activities of the past.

16. This is why we do not need different sets of rules by which to account for the political world and the scientific world. Similarly, we consider scientists' honesty and dishonesty from a single analytical perspective. Fraud and honesty are not fundamentally different kinds of behaviour; they are strategies whose relative value depends on the circumstances and the state of the agonistic field.

17. If reality means anything, it is that which "resists" (from the Latin "res"—thing) the pressure of a force. The argument between realists and relativists is exacerbated by the absence of an adequate definition of reality. It is possible that the following is sufficient: that which cannot be changed at will is what counts as real.

18. Although Brillouin is largely unknown among sociologists of science, he has made important contributions to a materialist analysis of science production. He regards *all* scientific activity (including the so-called "intellectual" or "cognitive" ones) as material operations which are in any way homologous to the usual object of physics. Since he provides a bridge between matter and information, he also bridges the gap—so dramatic for the study of science—between intellectual and material factors.

19. Even bench work can best be analyzed in terms of staging and writing. The samples are put into coloured racks on one side of the surgical table, and are moved slowly. The movement is monitored by a stop watch and recorded on a sheet of paper. Even at this level, possible objections are being countered by the set of precautions exercised in conducting this work (see Photograph File).

20. Many other aspects of the Go game analogy could be applied to the work of science. The main advantage of the analogy is that it provides an approximate illustration of the contingency/necessity dialectic. A further advantage is its illustration of the reification process in science. In Figure 6.1c, for example, the stone played at the fourth move lies next to another played at the 148th move. A group of white stones have been surrounded and are removed from the board. This approximates the movement of contradiction as shown in Chapter 3; whether or not a given formation is seen as contradictory (and requires elimination) will depend on the local context and on the pressures of the agonistic field. In this case, elimination will result from black's decision to play at a certain position.

21. One of the main interests of the field study is that the sociological work could be pursued hand-in-hand with the biological research of the institute. But it was clear to the observer that both his informants and his sociological colleagues were claiming to be doing science. The problems raised by this complicated relationship will be examined in detail elsewhere.

22. Our claim is not that we are advancing an original "paradigm" for the analysis of science. We simply aim to show how close our anthropological position is to other studies broadly named "sociology of science." Our impression is that the main approaches followed so far are (a) not connected to one another; (b) somewhat undecided on what is the final status of their findings. The slight, but radical, modification of background that we entertain here might provide a vantage point from which the importance of these findings can be fully appreciated.

23. This was due, in part, to the observer's isolation and lack of training and, in part, to the lack of any former anthropological studies of modern science. One particularly useful source was Auge's (1975) analysis of witchcraft in the Ivory Coast, which provides an intellectual framework for resistance to being impressed by scientific endeavour.

24. It seems that the basic prototype of scientific activity is not to be found in the realm of mathematics or logic but, as Nietzsche (1974) and Spinoza (1667) frequently pointed out, in the work of exegesis. Exegesis and hermeneutics are the tools around which the idea of scientific production has historically been forged. We claim that our empirical observations of laboratory activity fully support that audacious point of view; the notion of inscription, for example, is not to be taken lightly (Derrida, 1977).

25. "Fiction" is to be taken as having a noncommitted or "agnostic" meaning that can be applied to the whole process of fact production but to none of its stages in particular. The production of reality is what concerns us here, rather than any one produced final stage (stage 5 in the terminology of Ch. 2). Our main interest in using the word "fiction" is the connotation of literature and writing accounts. De Certeau once said (pers. com.), "There can only be a science of science-fiction." Our discussion is a first tentative step towards making clear the link between science and literature (Serres, 1977).

REFERENCES

Anonymous (1974) Sephadex: Gel Filtration in Theory and Practice. Uppsala: Pharmacia.

_____(1976a) B.L.'s interview. Oct. 19. Dallas.

_____(1976b) B.L.'s interview. Oct. 19. Dallas

ALTHUSSER, L. (1974) La philosophie spontanée des savants. Paris: Maspéro.

ARISTOTLE (1897) The Rhetorics of Aristotle. Trans. by E. M. Cope and J. E. Sandys. Cambridge.

ATLAN, H. (1972) L'organisation biologique et la théorie de l'information. Paris: Hermann.

AUGE, M. (1975) Théories des pouvoirs et idéologies. Paris: Hermann.

BACHELARD, G. (1934) Le nouvel esprit scientifique. Paris: P.U.F.

_____ (1953) Le matérialisme rationnel. Paris: P.U.F.

_____ (1967) La formation de l'esprit scientifique: contribution à une psychoanalyse de la connaissance approchée. Paris: Vrin.

BARNES, B. (1974) Scientific Knowledge and Sociological Theory. London: Routledge and Kegan Paul.

_____ and LAW, J. (1976) "Whatever should be done with indexical expressions?" Theory and Society 3 (2): 223-237.

BARNES, B. and SHAPIN, S. [eds.] (forthcoming) Natural Order: Historical Studies of Scientific Culture. Beverly Hills: Sage Publications.

BARTHES, R. (1957) Mythologies. Paris: Le Seuil.

_____ (1966) Critique et vérité. Paris: Le Seuil.

_____ (1973) Le plaisir du texte. Paris: Le Seuil.

BASTIDE, F. (forthcoming) Analyse sémiotique d'un article de science expérimentale. Urbino: Centre International de Sémiotique.

Beckman Instruments (1976) B.L.'s interview. Aug. 24. Palo Alto.

BEYNON, J. H. (1960) Mass Spectometry. Amsterdam: Elsevier.

BERNAL, J. D. (1939) The Social Function of Science. London: Routledge and Kegan Paul.

BERNARD, C. (1865) Introduction à l'étude de la Medicina Experimentale. Paris.

BHASKAR, R. (1975) A Realist Theory of Science. Atlantic Highlands, N. J.: Humanities Press.

BITZ, A., McALPINE, A., and WHITLEY, R. D. (1975) The Production, Flow and Use of Information in Research Laboratories in Different Sciences. Manchester Business School and Centre for Business Research.

BLACK, M. (1961) Models and Metaphors. Ithaca, N. Y.: Cornell University Press.

BLISSETT, M. (1972) Politics in Science. Boston: Little, Brown.

BLOOR, D. (1974) "Popper's mystification of objective knowledge." Science Studies 4: 65-76.

_____ (1976) Knowledge and Social Imagery. London: Routledge and Kegan Paul.

_____ (1978) "Polyhedra and the abominations of Leviticus." British Journal for the History of Science 11: 245-272.

BOGDANOVE, E. M. (1962) "Regulations of TSH secretion."Federations Proceeding 21: 623.

BOLER, J., ENZMANN, F., FOLKERS, K., BOWERS, C. Y., and SCHALLY, A. V. (1969) "The identity of clinical and hormonal properties of the thyrotropin releasing hormones and pyroglutamyl-histidine-proline amide." B. B. R. C. 37: 705.

BOURDIEU, P. (1972) Esquisse d'une théorie de la practique. Genève: Droz.

_____ ((1975a) "Le couturier et sa griffe." Actes de la Recherche en Sciences Sociales 1 (1).

_____(1975b) "The speficity of the scientific field and the social conditions of the progress of reason." Social Science Information 14 (6): 19-47.

_____ (1977) "La production de la croyance: contribution a une economie des biens symbolique." Actes de la Recherche en Sciences Sociales 13: 3-43.

BRAZEAU, P. and GUILLEMIN, R. (1974) "Somatostatin: newcomer from the hypothalamus." New England Journal of Medicine 290: 963-964.

BRILLOUIN, L. (1962) Science and Information Theory. New York: Academic Press.

_____ (1964) Scientific Uncertainty and Information. New York: Academic Press.

BROWN, P. M. (1973) High Pressure Liquid Chromatography. New York: Academic Press.

BULTMANN, R. (1921) Die Geschichte der synoptischen Tradition. Göttingen: Vandenhoek und Ruprecht. (Histoire de la tradition synoptique. Paris: Le Seuil (1973).)

BURGUS, R. (1976) B.L.'s interview. April 6. San Diego.

_____ and GUILLEMIN, R. (1970a) "Chemistry of thyrotropin releasing factor in hypophysiotropic hormones of the hypothalamus." Pp. 227-241 in J. Meites (eds.) Hypophysiotropic Hormones of the Hypothalamus. Baltimore: Williams and Wilkins.

_____ (1970b) "Hypothalamic releasing factors." Annual Review of Biochemistry 39: 499-526.

BURGUS, R., WARD, D. N. SAKIZ, E., and GUILLEMIN, R. (1966)"Actions des enzymes protéolytiques sur des préparations purifiées de l'hormone hypothalamique TSH (TRF)." C. R. de l'Ac. des sciences 262: 2643-2645.

BURGUS, R., DUNN, T. F., WARD, D. N., VALE, W., AMOSS, M., and GUILLEMIN, R. (1969a) "Dérivés polypeptidiques de synthèse doues d' activité hypophysiotrope TRF." C.R. de l'Ac. des Sciences 268: 2116-2118.

BURGUS, R., DUNN, T. F., DESIDERO, D., VALE, W., and GUILLEMIN, R. (1969b) "Dérivés polypeptidiques de synthèse doués d'activité hypophysiotrope TRF: nouvelles observations." C.R. de l'Ac. des Sciences 269: 226-228.

BURGUS, R., DUNN, T. F., DESIDERO, D., and GUILLEMIN, R. (1969c) "Structure moléculaire du facteur hypothalamique hypophysiotrope TRF d'origine ovine." C.R. de l'Ac. des Sciences 269: 1870-1873.

BURGUS, R., DUNN, T. F., DESIDERO, D., WARD, D. N., VALE, W., and GUILLEMIN, R. (1970) "Characterization of ovine hypothalamic TSH-releasing factor (TRF)." Nature 226 (5243): 321-325.

CALLON, M. (1975) "L'opération de traduction comme relation symbolique." In P. Roqueplo (ed.) Incidence des rapports sociaux sur le developpement scientifique et technique. Paris: C.N.R.S.

_____ (1978) De problèmes en problèmes: itinéraires d'un laboratoire universitaire saisi par l'aventure technologique. Paris: Cordes.

COLE, J. R. and COLE, S. (1973) Social Stratification in Science. Chicago: University of Chicago Press.

COLLINS, H. M. (1974) "The T.E.A. set: tacit knowledge and scientific networks." Science Studies 4: 165-186.

––––– (1975) "The seven sexes: a study in the sociology of a phenomenon or the replication of experiments in physics." Sociology 9 (2): 205-224.

––––– and COX, G. (1977) "Relativity revisited: Mrs. Keech—a suitable case for special treatment?" Social Studies of Science 7 (3) 372-381.

COSER, L. A. and ROSENBURG, B. [eds.] (1964) Sociological Theory. London: Macmillan.

COSTA de BEAUREGARD, O. (1963) Le second principe de la science du temps: entropie, information, irreversibilité. Paris: Le Seuil.

CRANE, D. (1969) "Social structure in a group of scientists: a test of the 'invisible college' hypothesis." American Sociological Review 34: 335-352.

–––––(1972) Invisible Colleges. London: University of Chicago Press.

––––– (1977) "Review symposium." Society for Social Studies of Science Newsletter 2 (4): 27-29.

CRICK, F. and WATSON, J. (1977) B.L.'s interview. Feb. 18. San Diego.

DAGOGNET, F. (1973) Ecriture et iconographie. Paris: Vrin.

DAVIS, M. S. (1971) "That's interesting." Philosophy of the Social Sciences 1: 309-344.

DE CERTEAU (1973) L'écriture de l'histoire. Paris: Le Seuil.

DERRIDA, J. (1977) Of Grammatology. Baltimore: Johns Hopkins University Press.

DONOVAN, B. T., McCANN, S. M., and MEITES, J. [eds.] (forthcoming) Pioneers in Neuroendocrinology, Vol. 2. New York: Plenum Press.

DUCROT, V. and TODOROV, T. (1972) Dictionaire encyclopédique des sciences du language. Paris: Le Seuil.

EDGE, D. O. [ed.] (1964) Experiment: A Series of Scientific Case Histories. London: BBC.

––––– (1976) "Quantitative measures of communication in science." Paper presented at the International Symposium on Quantitative Measures in the History of Science. Berkeley, California, Aug. 25-27.

––––– and MULKAY, M. J. (1976) Astronomy Transformed. London: Wiley-Interscience.

EGGERTON, F. N. [ed.] (1977) The History of American Ecology. New York: Arno Press.

FEYERABEND, P. (1975) Against Method. London: NLB.

FOLKERS, K., ENZMANN, F., BOLER, J. G., BOWERS, C. Y., and SCHALLY, A. V. (1969) "Discovery of modification of the synthetic tripeptide-sequence of the thyrotropin releasing hormone having activity." B.B.R.C. 37: 123.

FORMAN, P. (1971) "Weimar culture, causality and quantum theory 1918-1927." In Historical Studies in the Physical Sciences. Philadelphia: University of Pennsylvania Press.

FOUCAULT, M. (1966) Les mots et les choses. Paris: Gallimard.

––––– (1972) Histoire de la folie a l'age classique. Paris: Gallimard.

––––– (1975) Surveiller et punir. Paris: Gallimard.

––––– (1978) "Vérité et pouvoir." L'arc 70.

FRAME, J. D., NARIN, F., and CARPENTER, M. P. (1977) "The distribution of world science." Social Studies of Science 7: 501-516.

GARFINKEL, H. (1967) Studies in Ethnomethodology. Englewood Cliffs, N.J.: Prentice-Hall.

GARVEY, W. D. and GRIFFITH, B. C. (1967) "Scientific communication as a social system." Science 157: 1011-1016.

_____ (1971) "Scientific communication: its role in the conduct of research and creation of knowledge." American Psychologist 26: 349-362.

GELOTTE, B. and PORATH, J. (1967) "Gel filtration in chromotography." In E. Heftmann (ed.) Chromatography. New York: Van Nostrand Reinhold.

GILBERT, G. N. (1976) "The development of science and scientific knowledge: the case of radar meteor research." Pp. 187-204 in Lemaine et al. (eds.) Perspectives on the Emergence of Scientific Disciplines. The Hague: Mouton/Aldine.

GILPIN, R. and WRIGHT, C. [eds.] (1964) Scientists and National Policy Making. New York: Columbia University Press.

GLASER, B. and STRAUSS, A. (1968) The Discovery of Grounded Theory. London: Weidenfeld and Nicolson.

GOLDSMITH, M. and MACKAY, A. [eds.] (1964) The Science of Science. London: Souvenir.

GOPNIK, M. (1972) Linguistic Structure in Scientific Texts. Amsterdam: Mouton.

GREEP, R. O. (1963) "Synthesis and summary." Pp. 511-517 in Advances in Neuroendocrinology. Urbana: University of Illinois Press.

GREIMAS, A. J. (1976) Sémiotique et sciences sociales. Paris: Le Seuil.

GUILLEMIN, R. (1963) "Sur la nature des substances hypothalamiques qui controlent la sécrétion des hormones antéhypophysaires." Journal de Psysiologie 55: 7-44.

_____ (1975) B.L.'s interview. Nov. 28. San Diego.

_____ (1976) "The endocrinology of the neuron and the neural origin of endocrine cells." In J. C. Porter (ed.) Workshop on Peptide Releasing Hormones. New York: Plenum Press.

_____ and BURGUS, R. (1972) "The hormones of the hypothalamus." Scientific American 227(5): 24-33.

_____ SAKIZ, E., and WARD, D. N. (1966) "Nouvelles données sur la purification de l'hormone hypothalamique TSH hypophysiotrope, TRF." C. R. de l'Ac. des Sciences 262: 2278-2280.

GUILLEMIN, R., BURGUS, R., and VALE, W. (1968) "TSH releasing factor: an RF model study." Exerpta Medica Inter. Congress Series 184: 577-583.

GUILLEMIN, R., SAKIZ, E., and WARD, D. N. (1965) "Further purification of TSH releasing factor (TRF)." P.S.E.B.M. 118: 1132-1137.

GUILLEMIN, R., YAMAZAKI, E., JUTISZ, M., and SAKIZ, E. (1962) "Présence dans un extrait de tissus hypothalamiques d'une substance stimulant la sécrétion de l'hormone hypophysaire thyréotrope (TSH)." C. R. de l'Ac. des Sciences 255: 1018-1020.

GUSFIELD, J. (1976) "The literary rhetoric of science." American Sociological Review 41 (1): 16-34.

_____ (forthcoming) "Illusion of authority: rhetoric, ritual and metaphor in public actions—the case of alcohol and traffic safety."

HABERMAS, J. (1971) Knowledge and Human Interests. Boston: Beacon Press.

HAGSTROM, W. O. (1965) The Scientific Community. New York: Basic Books.

HARRIS, G. W. (1955) Neural Control of the Pituitary Gland. Baltimore: Williams and Wilkins.

_____ (1972) "Humours and hormones." Journal of Endocrinology 53: i-xxiii.

HARRIS, M. (1968) The Rise of Anthropological Theory. London: Routledge and Kegan Paul.

HEFTMANN, E. [ed.] (1967) Chromatography. New York: Van Nostrand Reinhold.

HESSE, M. (1966) Models and Analogies in Science. Notre Dame, IN: Notre Dame University Press.

HORTON, R. (1967) "African traditional thought and Western science." Africa 37: 50-71, 155-187.

HOYLE, F. (1975) Letter to the Times. April 8.

HUME, D. (1738) A Treatise of Human Nature. London.

JACOB, F. (1970) La logique du vivant. Paris: Gallimard.

_____ (1977) "Evolution and tinkering." Science 196(4295): 1161-1166.

JUTISZ, P., SAKIZ, E., YAMAZAKI, E., and GUILLEMIN, R. (1963) "Action des enzymes protéolytiques sur les facteurs hypothalamiques LRF et TRF." C.R. de la societe de Biologie 157 (2): 235.

KANT, E. (1950) [1787] Critique of Pure Reason. London: Macmillan.

KAWABATA, Y. (1972) The Master of Go. New York: Alfred A. Knopf.

KORACH, M. (1964) "The science of industry." Pp. 179-194 in Goldsmith and Mackay (eds.) The Science of Science. London: Souvenir.

KNORR, K. (1978) "Producing and reproducing knowledge: descriptive or constructive." Social Science Information 16(6): 669-696.

_____(forthcoming) "From scenes to scripts: on the relationships between research and publication in science."

_____ (forthcoming) "The research process: tinkering towards success or approximation of truth." Theory and Society.

KUHN, T. (1970) The Structure of Scientific Revolutions. Chicago: University of Chicago Press.

LACAN, J. (1966) Les écrits. Chapter: "La science et la vérité,"Pp. 865-879. Paris: Le Seuil.

LAKATOS, I. and MUSGRAVE, A. (1970) Criticism and the Growth of Knowledge. Cambridge: Cambridge Universtiy Press.

LATOUR, B. (1976) "Including citations counting in the systems of actions of scientific papers." Society for Social Studies of Science, 1st meeting, Ithaca, Cornell University.

_____ (1976b) "A simple model for a comprehensive sociology of science." (mimeographed).

_____ (1978, forthcoming) "The three little dinosaurs."

_____ and FABBRI, P. (1977) "Pouvoir et devoir dans un article de science exacte." Actes de la Recherche en Sciences Sociales 13: 81-95.

LATOUR, B. and RIVIER, J. (1977, forthcoming) "Sociology of a molecule."

LAW, J. (1973) "The development of specialities in science: the case of X-ray protein crystallography." Science Studies 3: 275-303.

LEATHERDALE (1974) The Role of Analogy, Model and Metaphor in Science. New York: Elsevier.

LECOURT, D. (1976) Lyssenko. Paris: Maspéro.

LEHNINGER, (1975) Biochemistry. New York: Worth.

LEMAINE, G., CLEMENÇON, M., GOMIS, A., POLLIN, B., and SALVO, B. (1977) Stratégies et choix dans la recherche apropos des travaux sur le sommeil. The Hague: Mouton.

LEMAINE, G., LÉCUYER, B. P., GOMIS, A., and BARTHÉLEMY, G. (1972) Les voies du succès. Paris: G.E.R.S.

LEMAINE, G., MACLEOD, R., MULKAY, M., and WEINGART, P. [eds.] (1976) Perspectives on the Emergence of Scientific Disciplines. The Hague: Mouton/ Aldine.

LEMAINE, G. and MATALON, B. (1969) "La lutte pour la vie dans la cité scientifique." Revue Française de Sociologie 10: 139-165.

LEVI-STRAUSS, C. (1962) La pensée sauvage. Paris: Plon.

LOVELL, B. (1973) Out of the Zenith. London: Oxford University Press.

LYOTARD, J. F. (1975, 1976) Lessons on Sophists. San Diego: University of California.

McCANN, S. M. (1976) B.L.'s interview. Oct. 19. Dallas.

MACHLUP, F. (1962) The Production and Distribution of Knowledge. Princeton, N.J.: Princeton University Press.

MANSFIELD, E. (1968) The Economics of Technological Change. New York: W.W. Norton.

MARX, K. (1970) Feuerbach: Opposition of the Naturalistic and Idealistic Outlook. New York: Beckman.

_____ (1977) The Capital, Vol. 1. New York: Random House.

MEDAWAR, P. (1964) "Is the scientific paper fraudulent? yes; it misrepresents scientific thought." Saturday Review Aug. 1: 42-43.

MEITES, J. [ed.] (1970) Hypophysiotropic Hormones of the Hypothalamus. Baltimore: Williams and Wilkins.

_____, DONOVAN, B., and McCANN, S. (1975) Pioneers in Neuroendocrinology. New York: Plenum Press.

MERRIFIELD, R.B. (1965) "Automated synthesis of peptides." Science 150 (8; Oct.): 178-189.

_____ (1968) "The automatic synthesis of proteins." Scientific American 218 (3): 56-74.

MITROFF, I. I. (1974) The Subjective Side of Science. New York: Elsevier.

MONOD, J. (1970) Le hasard et la nécessité. Paris: Le Seuil.

MOORE, S. (1975) "Lyman C. Craig: in memoriam." Pp. 5-16 in Peptides: Chemistry; Structure; Biology. Ann Arbor: Ann Arbor Science Publishers.

_____, SPACKMAN, D. H., and STEIN, W. H. (1958) "Automatic recording apparatus for use in the chromatography of amino acids." Federation Proceedings 17 (Nov.): 1107-1115.

MORIN, E. (1977) La méthode. Paris: Le Seuil.

MULKAY, M. J. (1969) "Some aspects of cultural growth in the natural sciences." Social Research 36(1): 22-52.

_____ (1972) The Social Process of Innovation. London: Macmillan.

_____ (1974) 'Conceptual displacement and migration in science: a prefatory paper." Social Studies of Science 4: 205-234.

_____ (1975) "Norms and ideology in science." Social Science Information 15 (4/5): 637-656.

_____, GILBERT, G. N., and WOOLGAR, S. (1975) "Problem areas and research networks in science." Sociology 9: 187-203.

MULLINS, N. C. (1972) "The development of a scientific specialty: the Phage group and origins of molecular biology." Minerva 10: 51-82.

_____ (1973) Theory and Theory Groups in Contemporary American Sociology. New York: Harper and Row.

_____ (1973b) "The development of specialties in social science: the case of ethnomethodology." Science Studies 3: 245-273.

NAIR, R.M.G., BARRETT, J. F., BOWERS, C. Y., and SCHALLY, A. V. (1970) "Structure of porcine thyrotropine releasing hormone." Biochemistry 9: 1103.

NIETZSCHE, F. (1974a) Human, All Too Human. New York: Gordon Press.

_____ (1974b) The Will to Power. New York: Gordon Press.

OLBY, R. (1974) The Path to the Double Helix. Seattle.: University of Washington Press.

ORGEL, L. E. (1973) The Origins of Life. New York: John Wiley.

PEDERSEN, K. O. (1974) "Svedberg and the early experiments: the ultra centrifuge." Fractious 1 (Beckman Instruments).

PLATO The Republic.

POINCARÉ, R. (1905) Science and Hypothesis. New York: Dover.

POPPER, K. (1961) The Logic of Scientific Discovery. New York: Basic Books.

PORATH, J. (9167) "The development of chromatography on molecular sieves." Laboratory Practice 16 (7).

PRICE, D. J. de SOLLA (1963) Little Science, Big Science. London: Columbia University Press.

_____ (1975) Science Since Babylon. London: Yale University Press.

RAVETZ, J. R. (1973) Scientific Knowledge and Its Social Problems. Harmondsworth: Penguin.

REIF, F. (1961) "The competitive world of the pure scientist." Science 134 (3494): 1957-1962.

RESCHER, N. (1978) Scientific Progress: A Philosophical Essay on the Economics of Research in Natural Science. Oxford: Blackwell.

RODGERS, R. C. (1974) Radio Immuno Assay Theory for Health Care Professionals. Hewlett Packard.

ROSE, H. and ROSE, J. [eds.] (1976) Ideology of/in the Natural Sciences. London: Macmillan.

RYLE, M. (1975) Letter to the Times, 4.12.

SACKS, H. (1972) "An initial investigation of the usability of conversational data for doing sociology." Pp. 31-74 in Sudnow (ed.) Studies in Social Interaction. New York: Free Press.

_____, SCHEGLOFF, E. A., and JEFFERSON, G. (1974) "A simplest systematics for the organisation of turn-taking for conversation." Language 50: 696-735.

SALOMON BAYET, C. (1978) L'institution de la science, et l'expérience du vivant." Ch. 10. Paris: Flammarion.

SARTRE, J. P. (1943) L'Etre et le Néant. Paris: Gallimard.

SCHALLY, A. V. (1976) B.L.'s interview. Oct. 21. New Orleans.

_____, AMIMURA, A., BOWERS, C. Y., KASTIN, A. J., SAWANO, S., and REDDING, T. W. (1968) "Hypothalamic neurohormones regulating anterior pituitary function." Recent Progress in Hormone Research 24: 497.

SCHALLY, A. V., ARIMURA, A., and KASTIN, A. J. (1973) "Hypothalamic regulatory hormones." Science 179 (Jan. 26): 341-350.

SCHALLY, A. V., BOWERS, C.Y., REDDING, T. W., and BARRETT, J. F. (1966) "Isolation of thyrotropin releasing factor TRF from porcine hypothalami." Biochem. Biophys. Res. Comm. 25: 165.

SCHALLY, A. V., REDDING, T. W., BOWERS, C. Y. and BARRETT, J. F. (1969) "Isolation and properties of porcine thryrotropin releasing hormone." J. Biol. Chem. 244: 4077.

SCHARRER, E. and SCHARRER, B. (1963) Neuroendocrinology. New York: Columbia University Press.

SCHUTZ, A. (1953) "The problem of rationality in the social world." Economica 10.

SERRES, M. (1972) L'interférence, Hermes II. Paris: Ed. de Minuit.

_____ (1977a) La distribution, Hermes IV. Paris: Ed. de Minuit.

_____ (1977b) La naissance de la physique dans la texte de Lucrèce: fleuves et turbulences. Paris: Ed. de Minuit.

SHAPIN, S. (forthcoming) "Homo Phrenologicus: anthropological perspectives on a historical problem." In Barnes and Shapin (eds.).

SILVERMAN, D. (1975) Reading Casteneda. London: Routledge and Kegan Paul.

SINGH, J. (1966) Information Theory, Language and Cybernetics. New York: Dover.

SOHN RETHEL, A. (1975) "Science as alienated consciousness." Radical Science Journal 2/3: 65-101.

SPACKMAN, N.D.H., STEIN, W. H., and MOORE, S. (1958) "Automatic recording apparatus for use in the chromatography of amino acids." Analytical Chemistry 30 (7): 1190-1206.

SPINOZA (1976) [1977] The Ethics. Appendix Part I. Secaucus, N.J.: Citadel Press.

SUDNOW, D. [ed.] (1972) Studies in Social Interaction. New York: Free Press.

SWATEZ, G. M. (1970) "The social organisation of a university laboratory." Minerva 8: 36-58.

TOBEY, R. (1977) "American grassland ecology 1895-1955: the life cycle of a professional research community." In Eggerton (ed.) The History of American Ecology. New York: Arno Press.

TUDOR, A. (1976) "Misunderstanding everyday life." Sociological Review 24: 479-503.

VALE, W. (1976) "Messengers from the brain." Science Year 1976. Chicago: F.E.E.C.

WADE, N. (1978) "Three lap race to Stockholm." New Scientist April 27, May 4, May 11.

WATANABA, S. [ed.] (1969) Methodologies of Pattern Recognition. New York: Academic Press.

WATKINS, J.W.N. (1964) "Confession is good for ideas." Pp. 64-70 in Edge (ed.) Experiment: A Series of Scientific Case Studies. London: BBC.

WATSON, J. D. (1968) The Double Helix. New York: Atheneum.

_____ (1976) Molecular Biology of the Gene. Menlo Park, CA: W. A. Benjamin.

WHITLEY, R. D. (1972) "Black boxism and the sociology of science: a discussion of the major developments in the field." Sociological Review Monograph 18: 61-92.

WILLIAMS, A.L. (1974) Introduction to Laboratory Chemistry: Organic and Biochemistry. Reading, MA: Addison-Wesley.

WILSON, B. [ed.] (1977) Rationality. Oxford: Blackwell.

WOOLGAR, S. W. (1976a) "Writing an intellectual history of scientific development: the use of discovery accounts." Social Studies of Science 6: 395-422.

_____ (1976b) "Problems and possibilities in the sociological analysis of scientists' accounts." Paper presented at 4S/ISA Conference on the Sociology of Science. Cornell, New York, Nov. 4-6.

_____ (1978) "The emergence and growth of research areas in science with special reference to research on pulsars." Ph.D. thesis, University of Cambridge.

WYNNE, B. (1976) "C. G. Barkla and the J phenomenon: a case study in the treatment of deviance in physics." Social Studies of Science 6: 307-347.

YALOW, R. S. and BERSON, S. A. (1971) "Introduction and general consideration." In E. Odell and O. Daughaday (eds.) Principles of Competitive Protein Binding Assays. Philadelphia: J. B. Lippincott.

YOUNG, B. (n. d.) "Science is social relations" (mimeographed).

POSTSCRIPT TO SECOND EDITION (1986)

There is a traditional tendency to chase and hound the "real" meaning of texts. Years after the initial publication of a volume, defenders and critics alike continue to argue over "what was actually intended" by its authors. As a welcome relief from this spectacle, literary theory has increasingly disavowed this kind of textual criticism. The current trend is to permit texts a life of their own. The "real" meaning of a text is recognised as an illusory or, at least, infinitely renegotiable concept. As a result, "what the text says," "what really happened" and "what the authors intended" are now very much up to the reader. It is the reader who writes the text.

Although this change has been most marked in the field of literary criticism, it clearly has a special relevance for the social study of science, which takes as axiomatic the tentative, contingent character of objectification practices. The construction of scientific facts, in particular, is a process of generating *texts* whose fate (status, value, utility, facticity) depends on their subsequent interpretation. In line with this notion of textual interpretation, we shall not attempt a definitive restatement of the argument of *Laboratory Life*, but instead offer comments on the nature of some of the criticisms of the book and on the changes in the social study of science which these criticisms reflect.

In early October 1975, one of us entered Professor Guillemin's laboratory for a two-year study of the Salk Institute. Professor Latour's knowledge of science was non-existent; his mastery of English was very poor; and he was completely unaware of the existence of the social studies of science. Apart from (or perhaps even because of) this last feature, he was thus in the classic position of the ethnographer sent to a completely foreign environment. Since the question has often been asked, it is useful to begin with a few words about how he got to the Salk Institute in the first place.

While in the Ivory Coast, as a researcher in the sociology of development with the French research institution ORSTOM, he had been asked to explain why it was so difficult for black executives to adapt to modern industrial life (Latour, 1973). He found a vast literature on African philosophy and in comparative anthropology. Right from the start, however, it seemed that many features were attributed a little too quickly to the African "mind," and that these could be more simply explained by social factors. For example, the young boys in technical schools were accused by their white teachers of being unable to "see in three dimensions." This was regarded as a serious deficiency. It turned out, however, that the school system (an exact replica of the French system) introduced engineering drawing to its pupils before they did any practical work on engines. Since the boys mostly came from country districts and had never seen or handled an engine before, the interpretation of the drawings presented them with quite a puzzle. As the study proceeded, the established preference for far-fetched cognitive explanations over simpler social ones became more evident. A terrible doubt arose: perhaps the entire literature on cognitive abilities was fundamentally wrong. It was especially troubling that every study depended on a distinction between scientific and prescientific reasoning. Stimulated by interaction with remarkable anthropologists like Marc Augé and other colleagues at ORSTOM, a rudimentary research programme took shape. What would happen to the Great Divide between scientific and prescientific reasoning if the same field methods used to study Ivory Coast farmers were applied to first-rate scientists? Two years before, the would-be anthropologist of science had met Professor Guillemin (like him, a native of Burgundy). Guillemin praised the openness of the Salk Institute and had invited him to carry out an epistemological study of his laboratory, providing he secured his own source of funding. It is worth acknowledging Guillemin's unusual generosity in providing total access to his laboratory and his forbearance in taking in (someone he took to be) an "epistemologist" (Dr. Jekyll) who subsequently turned into a sociologist of science (Mr. Hyde).[1]

When the first edition of *Laboratory Life* appeared in 1979, it was surprising to realise that this was the first attempt at a detailed study of the daily activities of scientists in their natural habitat. The scientists in the laboratory were probably more surprised than anyone that this was the only study of its kind. To them, our arguments about the need for such studies were obvious. "How could anyone *ignore* the details of our daily work?" they quipped. So their main reaction to the book (apart from carefully scrutinising the pseudonymous quotations we used) was that it was all rather unsurprising, if not trivial. Although this reaction is a nice confirmation of the accuracy of our observations, this is not our point. The scientists were much more attracted to Wade's subsequent (1981) rendering of the Guillemin-Schally controversies. Wade's book is an interesting, although very one-sided (pro Schally) account, but its chief value is in demonstrating the difference between good scientific journalism and the sociological study of science. Wade's sense of outrage is evident throughout as he delightedly portrays the ways in which the "rules of scientific method" were broken. The very episodes we avoided because they were too obviously "social" (in the limited sense of scandal-mongering and making "cheap shots") are highlighted by Wade's book. The prolongation of the dispute in Wade's account better approximated the scientists' more prurient interests than did ours. Clearly, our book was intended for a rather different audience.

Within the general community of scholars interested in theories of science, the novelty of the book's approach was also something of a surprise. Despite subsequent modifications (Kuhn, 1970) of his position, Kuhn (1962) had already provided (although perhaps unwittingly—see Kuhn, 1984) the general basis for a conception of the social character of science, and Barnes (1974) and Bloor (1976) had set the agenda for a "strong programme" in the sociology of scientific knowledge. The antipathy of many writers to the treatment of science as a "black box" was well established. We might thus have expected participant observation studies to be an integral feature of the wave of neo-Kuhnian analyses which characterised the sociology of science in the 1970s. But calls for full-blooded sociology of science were not immediately met by participant observation studies. Few had spent a significant amount of time in close proximity with the day-to-day activities of working scientists.[2]

With the benefit of hindsight, it is possible to place this initial failure of nerve in perspective. Of course, any ethnographer (or participant observer) will testify to the taxing demands of entering and living in a strange culture. The esoteric culture of the scientific laboratory provides particularly daunting problems, both conceptual and practical. For example, the problem of maintaining analytic distance is acute for the ethnographer of science because his own (native) culture is itself infused with notions of what science is like. More significant, perhaps, is the fact that the sociology of science of the late 1970s responded rather slowly to some of the implications of Kuhn's work. It is well known that Kuhn's work coincided with a fundamental re-evaluation of preconceptions about the "special" character of science, one particular consequence of which was a change of focus in the social study of science. Instead of studying relationships between scientists, the reward system and institutional affiliations, the trend was to demonstrate the fundamentally social character of the objects, facts and discoveries of science. The sociology of science became a sociology of scientific knowledge.

Less well realised, perhaps, is that the same re-evaluation of preconceptions about science also has implications for the methods and techniques adopted by the social study of science. The revision of epistemological preconceptions about science raises awkward questions about the nature of its social analysis. Can we go on being instrumentally realist in our own research practices while proclaiming the need to demystify this tendency among natural scientists? Should we be vocal about the social processes of science, hitherto hidden from view, and yet silent about the social processes of our own research? The hesitant, differential response to this deep-rooted issue partly accounts for the proliferation of research perspectives which has accompanied the release from pre-Kuhnian orthodoxy. Although generally united in their disdain for the traditional ("received") view of science, practitioners of the new social study of scientific knowledge differ markedly in their methodological styles and preferences (see for example the collection of papers in Knorr-Cetina and Mulkay, 1983). The differential response to the revision of epistemological preconceptions also begins to account for the variation of responses to *Laboratory Life*.[3]

A general complaint about the book concerned its indiscipline. One reviewer remarked that reading *Laboratory Life* was "rather like taking an extremely bumpy ride over fascinating terrain" (Westrum, 1982: 438). Apart from noting the omission of a detailed table of contents and the absence of an index (both faults rectified in this edition), Westrum speaks of a lack of a sense of unity, the lack of continuous action and the relative incoherence of the narrative. But our aim was precisely to avoid giving the kind of smoothed narrative characteristic of traditional constructions of the "way things are." For example, we did not want an account in which the early presentation of *dramatis personnae* implied that humans were to be taken as the primary category of actors within the laboratory. Westrum himself notes the congruence between the form of our report and the process we describe in the laboratory: "Like the animal brains which get chopped up in the course of the researchers' work, the human struggle of the researchers to advance science and their own careers is chopped up so that Latour and Woolgar can examine and classify the interactions between them" (Westrum, 1982: 438). A more prosaic explanation of the form of *Laboratory Life* stems from the nature of collaboration between a French philosopher and a British sociologist. In the best tradition of innovation through hybridisation, the authors found themselves continually rediscovering and renegotiating the significance of the cultural divide known (chauvinistically) as the English Channel. From this process emerged an uneasy (but evidently fruitful) compromise of styles.

The more substantive criticisms of the book range over a variety of issues, of which the most important are summarised below. Rather than using space to develop a full scale rebuttal of each point, we offer brief comments about their significance and the problems they pose for future work.

How Radical is Radical?

While enthusiastically welcoming the detailed empirical demonstration of the argument that no part of science is beyond sociological analysis, some marxist scholars are critical that *Laboratory Life* is ultimately the product of a "bourgeois sociology of science" (Stewart, 1982: 133). Given the demonstration that scientific facts are formulated in the denial and obliteration of their own historicity, given that the internal relations of science can be described as distinctively capitalist, these critics are disappointed that we did not go on to ask why this is so. They complain that we fail to examine the connection between the construction of scientific facts and the hierarchical and exploitative relations of science and the class divisions of society as a whole. *Laboratory Life* is accused of espousing an idealist relativism whereby the absence of socio-economic analysis reduces material reality to "the almost arbitrary vicissitudes of human subjectivity" (Stewart, 1982: 135).

"Relativism" is evidently the bogey of this particular brand of radicalism. Indeed, there is a tendency to use it indiscriminately, for example, to fail to appreciate the distinction between relativism and constructivism. But the weakness of marxist analyses of science is their desire for a scientific/objective point of view. The proponents of marxist analyses of science need a critique of objectivity so as to make room for their radical science, but they also want a "real science" with which to ground this radical science (Latour, 1982a: 137; see also Wolff, 1981). The call for a macro-social analysis of the way in which the social relations of production lead scientists "to select from and shape nature in a particular way" (Stewart, 1982: 135) claims for marxist science the very privileges it denies to bourgeois science.

What Does It Mean to be Ethnographic?

The idea of an ethnographic study of scientific practice has given rise to a body of work which has come to be called "laboratory studies."[4] The common assumption of these studies is that our understanding of science can profitably draw upon experiences gained while immersed in the day-to-day activities of working scientists. However, beyond this there is little consensus as to what can and should be made of these experiences. In *Laboratory Life*, we indicated that our use of the term "anthropology of science" was intended to denote the presentation of preliminary empirical material, our desire to retrieve something of the craft character of science, the necessity to bracket our familiarity with the object of study, and our desire to incorporate a degree of "reflexivity" into our analysis. Now these features only correspond to the requirements of traditional ethnography in a rather general way. To call something ethnographic traditionally requires that we include a description of the tribe's ecology, technology and belief systems. But, as Knorr-Cetina (1982a: 40) has noted, this particular interpretation of "ethnography" has been heavily criticised within anthropology. A more general interpretation of the call for ethnography denotes the need for detailed empirical observations and field notes, especially where these include information about sources of funding, the career backgrounds of participants, the citation patterns in the relevant literature, the nature and origins of instrumentation and so on. These are necessary, in one view (Latour, 1982b), if we are to proceed to a comparative analysis of the local settings of fact production. An alternative view is that such details are necessary, not so much for comparative purposes, but because any attempt to grapple with the problems of describing science proceeds best from an empirical base.

In our original use of the term, we particularly stressed the utility of an "ethnographic" approach for maintaining analytic distance upon explanations of activity prevalent within the culture being observed. In the case of a scientific culture in particular, there is a strong tendency for the objects of that culture (facts) to provide their own explanation. Rather than produce an account which explained scientists' activities in terms of the facts which they discovered, our interest was to determine how a fact came to acquire its character in the first place. Lynch (1982) points out that our strategy corresponds to Schutz's (1944) recommendation that sociology adopt the perspective of the stranger, whereby the problems of making sense of an alien culture provide insights into those aspects of culture taken for granted by its members.

Lynch notes, as we ourselves do, that the technical practices of laboratory science involve the assessment of the relationship between "objective" and "socio-historical" states of affairs (Lynch, 1985b). However, Lynch stresses that this assessment by scientists (which he calls endogenous critical inquiry) operates independently of any professional sociological interest and does not rely solely upon approved social science methods (Lynch, 1982: 501). By contrast, the efforts of the social scientist, of which the use of the stranger device is one example, draw upon the stranger's own analytic competences as a social scientist. As a result, says Lynch, our use of anthropological strangeness produces a "disengaged" analysis which severs "the transitivity of technical practices to their real-worldly objects of study" (Lynch, 1982: 503).

What is the sense of "disengaged" in Lynch's usage? For Lynch, the competences of the sociologist are fundamentally distinct from those of the scientist, and the relationship between the two is problematic. As evidence of this distinctiveness, Lynch cites the failure of our "observer" (see especially Chapter 2) to perform the practices of the laboratory competently, as well as disputes between the observer and his informants, his lack of understanding of the technical reports and so on. Lynch argues that whatever distinguishes social scientific practices from those of the scientist is yet to be discovered.

Lynch's criticism depends both on a rigid distinction between insider (scientist) and outsider (observer) and upon a rather idealistic notion about the possibility of distinctively assigning competences to these categories. Whereas we began by wishing to avoid this distinction i.e. by not wanting to presume the principled difference between scientist and non-scientist, Lynch points out that the stranger device entails our use of this distinction. Lynch himself assumes the difference and complains that our reports of the observer's experiences exemplify a failure adequately to document the practices of the scientist. Lynch displays a commitment to an (actual) objective character of the technical practices and the real worldly objects of study. Although his criticism is a salient warning against sociology-centrism, it is unclear what would count as an adequately "engaged" account of scientists' technical practices.[5]

Our current position on the notion of "ethnography" is slightly different. Its main advantage is that unlike many kinds of sociology (especially marxist), the anthropologist *does not know* the nature of the society under study, nor where to draw the boundaries between the realms of technical, social, scientific, natural and so on. This additional freedom in defining the nature of the laboratory counts for much more than the artificial distance which one takes with the observed. This kind of anthropological approach can be used on any occasion when the composition of the society under study is uncertain. It is not necessary to travel to foreign countries to obtain this effect, even though this is the only way that many anthropologists have been able to achieve "distance." Indeed, this approach may very well be compatible with a close collaboration with the scientists and engineers under study. We retain from "ethnography" the working principle of *uncertainty* rather than the notion of exoticism.

The Place of Philosophy

It is part of the folk wisdom of the field that historians have been increasingly enthusiastic about new developments in the sociology of scientific knowledge while philosophers of science have remained more resistant. Certainly, there has been a marked antipathy to some forms of philosophy by sociologists. Philosophy bashing perhaps reached its apogee with Bloor's (1976: 45) comment that "to ask questions of the sort which philosophers address to themselves is usually to paralyse the mind." Since the debate between Bloor (1981) and Laudan (1981), however, some philosophers have evinced sympathy for the work of the sociology of scientific knowledge (for example, Nickles, 1982, 1984). This suggests that it is perhaps no longer productive to dismiss all attempts at philosophising science (Knorr-Cetina, 1982a).

One good reason for not dismissing philosophy is that the positions of most authors both within and beyond the social study of science are based on deep-seated ontological commitments rather than upon any empirical account of science. This is why empirical evidence (of the sort provided by *Laboratory Life*) is unlikely to change any minds. And this is why those who read the book through realist spectacles will see error (for example, Bazerman, 1980: 17). It is instead necessary to examine the very roots of these ontologies and to attempt to develop an alternative (Latour, 1984, 1986a). However, the particular branch of philosophy—epistemology—which holds that the only source of knowledge are ideas of reason intrinsic to the mind, is an area whose total extinction is overdue. The redundancy of epistemology is well established by flourishing sociological, historical and (other) philosophical analyses of knowledge, despite its constant assertion (directed in particular at the work of Bachelard and his French followers) of the impossibility of these disciplines. It is not that we need to apportion subject matter between epistemology and naturalistic studies of science and technology; the work of the latter is a dissipation of the former. So *Laboratory Life* is neither an attempt to develop an alternative epistemology nor is it an attack on philosophy. Perhaps the best way to express our position is by proposing a ten-year moratorium on cognitive explanations of science. If our French epistemologist colleagues are sufficiently confident in the paramount importance of cognitive phenomena for understanding science, they will accept the challenge. We hereby promise that if anything remains to be explained at the end of this period, we too will turn to the mind!

Perhaps the most interesting (philosophical) interpretation of our work is an attempt to enroll *Laboratory Life* as a confirmation (!) of the falsificationist theory of science. In this view, *Laboratory Life* constitutes "a striking corroboration" of Popperian philosophy of science (Tilley, 1981: 118): (our description of) the amount of effort invested by scientists in undermining each others' claims is the best proof that science is fundamentally different from every day common sense. Debates in everyday life are not settled by using huge laboratories and carefully staged controversies.

Tilley's boomerang (but nonetheless plausible) interpretation of our argument is useful because it reveals two basic flaws in our work. First, although it was originally both necessary and desirable, the laboratory should not be studied as an isolated unit; it is only one part of a wider story. The other part examines the way in which a laboratory becomes an obligatory reference point in all discussions. Not until the inner workings of the laboratory are studied in conjunction with the strategic positioning of the laboratory in society can Tilley's kidnapping manoeuvre be resisted. The full story will establish that there is a continuum between controversies in daily life and those occurring in the laboratory, and the investigation of this continuum will explain why more resources are needed in a laboratory than is usually necessary in a pub (Latour, 1986a and b). Second, Tilley demonstrates that the resources at our disposal are insufficient to force our particular interpretation in preference to any other. At almost no cost, Tilley has been able to produce a diammetrically opposed interpretation of the one we intended (see p. 284).

The Demise of the "Social"

A misunderstanding which has been more consequential with the expansion of social studies of science, concerns the use of the word "social." Given our explicit disavowal of "social factors" in the first chapter, it is clear that our continued use of the term was ironic. So what does it mean to talk about "social" construction? There is no shame in admitting that the term no longer has any meaning. "Social" retained meaning when used by Mertonians to define a realm of study which excluded consideration of "scientific" content. It also had meaning in the Edinburgh school's attempts to explain the technical content of science (by contrast with internalist explanations of technical content). In all such uses, "social" was primarily a term of antagonism, one part of a binary opposition. But how useful is it once we accept that *all* interactions are social? What does the term "social" convey when it refers equally to a pen's inscription on graph paper, to the construction of a text and to the gradual elaboration of an amino-acid chain? Not a lot. By demonstrating its pervasive applicability, the social study of science has rendered "social" devoid of any meaning (cf. Latour 1986a and b). Although this was also our original intention, it was not clear until now that we could simply ditch the term: our new subtitle now denotes our interest in "the construction of scientific facts."

Reflexivity

We earlier noted that one of our original concerns was to produce an "ethnographic" study which incorporated a degree of reflexivity. We also suggested that the diversity of reaction to *Laboratory Life* corresponded to a deep-seated ambivalence about the character and status of work in the social study of science, especially where this recognises itself as the construction of fictions about fiction construction. It is interesting, however, that most "laboratory studies" tended to adopt an instrumental rather than reflexive conception of ethnography (Woolgar, 1982). The current programmatic slogan of many laboratory studies (also common to much social study of science more generally) is the injunction to study science *as it happens*. In one sense of this clause, the work of laboratory studies is an attempt to produce a description of scientific work relatively unhindered by retrospective reconstruction: contemporaneous monitoring of scientific activity enables the analyst to base discussion on first-hand experiences rather than to rely on recollections made in the light of subsequent events. In a second sense, the study of science as it happens enables the analyst to bypass intermediary constructions arising from reliance on informants in situations removed from their everyday working environment. Thus, *in situ* observation provides more direct access to events in the laboratory than, for example, interview responses. In both cases, the general idea is that more is to be gained from being on the spot than from attempting interpretation from a secondary perspective. The *in situ* monitoring of contemporaneous scientific activity portrays the scientist located firmly at the laboratory bench and treats with some scepticism the kind of representations provided by the scientist, especially where these are produced in situations removed (either temporally or contextually) from the scene of the scientific action.

The straightforward interpretation of the "as it happens" clause implies that laboratory studies yield a "better" or "more accurate" picture of science than those studies relying on "distorted" versions preferred by actors removed from the scene. Undoubtedly, this line of argument is of some value, for example, in negotiating access to laboratories. Some scientists attach considerable weight to a contrast between, for example, "what philosophers like Popper say about science" and "what actually goes on in science." Yet the adoption of this line of argument for *analytic* purposes is both arrogant and entirely misleading. It presumes privileged access to the "real truth" about science and it suggests that this truth will eventually emerge from closer and more detailed observation of technical practices (cf. Gieryn, 1982). It thereby ignores the very phenomenon in need of investigation—namely the ways in which descriptions and reports of observations are variously presented (and received) as "good enough," "inadequate," "distorted," "real," "accurate" and so on.

A more reflexive appreciation of laboratory studies is less dismissive of what might be called "the problem of fallibility": the argument that *all* forms of description, report, observation and so on can always be undermined. However, instead of using this argument ironically (Woolgar, 1983), as a way of characterising the work of *others* (scientists or other sociologists) while implying that our own recommended alternative is free from such deficiencies, we should accept the universal applicability of fallibility and find ways of coming to terms with it. Instead of utilizing it in a merely critical role, the aim would be to retain and constantly draw attention to the phenomenon in the course of description and analysis. We might as well admit that as a "problem" it is both insoluble and unavoidable, and that even efforts to examine *how* it is avoided are doomed in that they entail an attempt to avoid it.[7] We need to explore forms of literary expressive whereby the monster can be simultaneously kept at bay and allowed a position at the heart of our enterprise.[8]

Of course, one interesting aspect of the exploration of reflexivity is that our writing is conventionally constrained by the use of report-like formats. This increases the tendency for ethnographies to be read as straightforwardly reporting on the "actual" state of affairs to be found in the laboratory. This kind of reading is not without value. Some will discover in this reading aspects of the world of scientific work of which they were previously unaware. But such reading misses the point. We attempted (especially in Chapter 2) to address the issue of reflexivity by placing the burden of observational experience on the shoulders of a mythical "observer." We attempted to alert the reader to the nature of his relationship with the text (and by implication to the nature of readers' relationships with all attempts to constitute objectivities through textual expression). For example, Photograph 1 (p. 93) is labelled "View from the laboratory roof." Now, presumably, a determinedly instrumentally minded reader will take this at face value, and go away happy that he is better informed about the character of laboratory roofs (and views therefrom). For such readers we are naturally pleased to increase their sum of knowledge about the world. But, unfortunately, much would have been lost. We hoped that the inclusion of such a photograph might at least make such readers stop and think about what is involved in the juxtaposition of textual imagery, and how this affects the reader's relationship with the "facts" as represented by the text. Our concerns for reflexivity would perhaps have begun to succeed where the text suggests to the reader that he ask himself whether or not the observations really took place, whether or not Jonas Salk really wrote the introduction, and so on.

Reflexivity is thus a way of reminding the reader that *all* texts are stories. This applies as much to the facts of our scientists as to the fictions "through which" we display their work. The story like quality of texts denotes the essential uncertainty of their interpretation: the reader can never "know for sure." We mentioned already the value of ethnography in stressing this uncertainty. Here we see that reflexivity is the ethnographer of the text.

Conclusion

The concluding chapter of *Laboratory Life* addresses the status of our own account, the question of whether or not we are (merely) supplying a new fiction (about science) for old. In the closing section of the original draft we declared that our analysis was "ultimately unconvincing." We asked readers of the text not to take its contents seriously. But our original publishers insisted that we remove the sentence because, they said, they were not in the habit of publishing anything that "proclaimed its own worthlessness."

It should be clear that we never claimed that our account is more privileged than those of our scientist-informants, nor that it is immune from criticism. But this statement, like the sentence excluded from the original draft, has often been interpreted as self-defeating: how could we not believe in our own account? How can we relativise both natural sciences and our own relativistic story? Clearly, readers can miss the *point* of the reflexivity and hear only apology and self contradiction. But the statement is only an *aporia* from the point of view of those who believe in the intrinsic existence of accurate and fictitious accounts per se. And this is precisely the point of view we dispute. For this reason, the final sentence of the first edition (which reappears as the last sentence of this postscript) tries to anticipate the amount of work necessary to make our interpretation more likely than others. It is a reminder that the value and status of any text (construction, fact, claim, story, this account) depend on more than its supposedly "inherent" qualities. As we suggested earlier, the degree of accuracy (or fiction) of an account depends on what is subsequently made of the story, not on the story itself. This is the fundamental principle we showed to be a work in the modalising and demodalising of statements. *Laboratory Life* is once again in the hands of its readers, exactly like TRF, TRH, Somatostatin and the other fact(or)s we discussed. It is others who transform the status of these claims, making them more or less factual, dismembering them, incorporating them into black-boxes for different argumentative purposes, ridiculing them and so on. There is neither self-contradiction nor self-defeat in recognising this common destiny of all claims. On the contrary, once the common fate of claims is recognised, it becomes easier to understand the differences in predicting each reader's behaviour. Each text, laboratory, author and discipline strives to establish a world in which its own interpretation is made more likely by virtue of the increasing number of people from whom it extracts compliance. In other words, interpretations do not so much *in*form as *per*form. From this perspective, our scientists are obviously better equipped at performing the world we live in than we are at deconstructing it. The recognition of this vast difference is in no way self-defeating. It merely acknowledges the present balance of forces. How much further research, investment, redefinition of the field, and transformation of what counts as an acceptable argument are necessary to make this account more plausible than its alternatives?

NOTES

1. Which is the bad guy? Readers are invited to transpose the identities here. The important point is that a metamorphosis occurred.

2. Fleck's (1979) account of the way in which the Wasserman reaction came to be related to syphillis is now widely acknowledged to have predated this trend, having been originally published in German in 1935. Westrum (1982) has suggested that Perry's (1966) study of psychiatric research also anticipates the conclusions of *Laboratory Life*.

3. Reviews and review articles which discuss *Laboratory Life* are listed in Additional References (p. 287), denoted by an asterisk (*). Many "laboratory studies" (see note 4) also include a critical evaluation of *Laboratory Life*.

4. For reviews of the field of "laboratory studies" see, for example, Knorr-Cetina, 1983; Woolgar, 1982. Empirical analyses falling under the "laboratory studies" rubric include investigations of the following substantive areas: neuroendocrinology (Latour and Woolgar, 1979; Latour, 1980, 1981), plant protein research (Knorr, 1977, 1979; Knorr-Cetina, 1981, 1982a, 1982b), brain science (Lynch, 1982, 1985a, 1985b), psychophysiology (Star, 1983), particle physics (Traweek, 1980, 1981, forthcoming), solid state physics (Woolgar, 1981a, 1981b, forthcoming), colloid chemistry (Zenzen and Restivo, 1982), catalytic chemistry (Boardman, 1980), cell biology (Law and Williams, 1981, 1982; Williams and Law, 1980), wildlife biology (McKegney, 1982) and limnology (Grenier, 1982, 1983). In addition, a number of articles provide general discussions of the importance of "anthropological approaches" to science (Anderson, 1981; Elkana, 1981; Lepenies, 1981) but tend neither to refer to, nor make use of, specific empirical work. A further detailed study of an individual scientist's experiences, but which fails to address the social process of laboratory work, is Goodfield (1981).

5. In a marvellously detailed account of the work of a neurosciences laboratory, Lynch himself does not claim to have achieved more than "the most speculative grasp of neuroscientific work within the monstrously difficult strictures of Garfinkel's program" (Lynch, 1985b: 128).

6. A recent meeting of the American Association for the Advancement of Science (New York, Summer 1984) included a session entitled "Laboratory Studies: What Scientists *Really* Do."

7. The "linguistic turn" in science studies can be glossed as the attempt to topicalise the ways in which scientists themselves do the work of interpretation despite the problem of fallibility. For example, the focus of "discourse analysis" is upon scientists' organisation of meaning, given the interpretive flexibility of, and variation between, their accounts (for example, Mulkay et al., 1983; Gilbert and Mulkay, 1984). Such studies fail the requirements of reflexivity to the extent that they purport to reveal (non-ironically) the actual discourse practices of scientists. For a general review of the large volume of different approaches to scientific texts, see Callon et al., 1986.

8. Some recent attempts to pursue this line are Ashmore (1985), Mulkay (1984) and Woolgar (1984).

ADDITIONAL REFERENCES

* denotes review (or review article) of *Laboratory Life*

ANDERSON, R. S. (1981) "The necessity of field methods in the study of scientific research." Pp. 213-244 in Mendelsohn and Elkana (1981).
ASHMORE, MALCOLM (1985) A Question of Reflexivity: Wrighting Sociology of Scientific Knowledge. Unpublished Ph.D. thesis. University of York.
* AUSTIN, J. (1982) Social Science and Medicine 16: 931-934.
* BAZERMAN, CHARLES (1980) 4S Newsletter 5: 14-19.
* BEARMAN, DAVID (1979) *Science 206*: 824-825.
BERGER, P. L. and LUCKMAN, T. (1971) The Social Construction of Reality. Harmondsworth: Penguin.
BLOOR, DAVID (1981) "The strengths of the strong programme." Philosophy of the Social Sciences 11: 173-198.
BOARDMAN, M. (1980) The Sociology of Science and Laboratory Research Practice: Some New Perspectives in the Social Construction of Scientific Knowledge. B.Sc. dissertation, Brunel University.
BORGES, J. L. (1981) "Pierre Menard, author of The Quixote." Pp. 96-103 in Borges: A Reader. Ed. E. R. Monegal and A. Reid. New York: E. P. Dutton.
CALLON, MICHEL, LAW, JOHN and RIP, ARIE [eds.] (1986) Texts and Their Powers. London: Macmillan.
* COZZENS, SUSAN (1980) 4S Newsletter 5: 19-21.
ELKANA, YEHUDA (1981) "A programmatic attempt at an anthropology of knowledge." Pp. 1-76 in Mendelshon and Elkana (1981).
FLECK, LUDWIG (1979) The Genesis and Development of a Scientific Fact, translation by F. Bradley and T. J. Trenn of 1935 German edition. Chicago: The University of Chicago Press.
GIERYN, T. (1982) "Relativist/constructivist programmes in the sociology of science: redundance and retreat." Social Studies of Science 12: 279-297.
GILBERT, G. NIGEL and MULKAY, MICHAEL (1984) Opening Pandora's Box. Cambridge: Cambridge University Press.
GOODFIELD, JUNE (1981) An Imagined World. New York: Harper and Row.
GRENIER, M. (1982) Toward An Understanding of the Role of Social Cognition in Scientific Inquiry: Investigations in a Limnology Laboratory. M.A. dissertation, McGill University.
———— (1983) "Cognition and social construction in laboratory science." 4S Review 1 (3): 2-16.
*HARAWAY, D. (1980) Isis 71: 488-489.
KNORR, K. D. (1977) "Producing and reproducing knowledge: descriptive or constructive? Towards a model of research production." Social Science Information 16: 699-696.
———— (1979) "Tinkering toward success: prelude to a theory of scientific practice." Theory and Society 8: 347-376.
———— and KROHN, R. and WHITLEY, R. D. [eds.] (1980) The Social Process of Scientific Investigation. Sociology of the Sciences Yearbook, Vol. 4. Dordrecht and Boston, Mass.: Reidel.

KNORR-CETINA, KARIN D. (1981) The Manufacture of Knowledge: An Essay on the Constructivist and Contextual Nature of Science. Oxford: Pergammon.
——— (1982a) "Reply to my critics." Society for the Social Studies of Science Newsletter 7 (4): 40-48.
——— (1982b) "Scientific communities or transepistemic arenas of research? A critique of quasi-economic models of science." Social Studies of Science 12: 101-130.
——— (1983) "The ethnographic study of scientific work: towards a constructivist interpretation of science." Pp. 116-140 in Knorr-Cetina and Mulkay (1983).
——— and MULKAY, MICHAEL (1983) [eds.] Science Observed: Perspectives on the Social Study of Science. London: Sage.
KUHN, T. S. (1962) The Structure of Scientific Revolutions. Chicago: University of Chicago Press.
——— (1970) The Structure of Scientific Revolutions. Second Edition, Enlarged. Chicago: University of Chicago Press.
——— (1984) "Reflections on receiving the John Desmond Bernal Award." 4S Review 1 (4): 26-30.
* KROHN, R. (1981) Contemporary Sociology 10: 433-434.
LAUDAN, LARRY (1981) "The pseudo-science of science?" Philosophy of the Social Sciences 11: 173-198.
LATOUR, BRUNO (1973) "Les idéologies de la competence en milieu industriel à Abidjan." Cahiers Orstrom Sciences Humaines 9: 1-174.
——— (1980) "Is it possible to reconstruct the research process? The sociology of a brain peptide." Pp. 53-73 in Knorr et al. (1980).
——— (1981) "Who is agnostic or What could it mean to study science?" in H. Kuklick and R. Jones (eds.), Knowledge and Society: Research in Sociology of Knowledge, Sciences and Art. London: JAI Press.
——— (1982a) "Reply to John Stewart," Radical Science Journal 12: 137-140.
——— (1982b) Review of Karin Knorr-Cetina's The Manufacture of Knowledge. Society for Social Studies of Science Newsletter 7 (4): 30-34.
——— (1983) "Give me a laboratory and I will raise the world." Pp. 141-170 in Knorr-Cetina and Mulkay (1983)
——— (1984) Les Microbes: Guerre et Paix suivi par Irréductions. Paris: A. M. Metailie et Pandore.
——— (1986a) The Pasteurisation of French Society, translated by Alan Sheridan. Cambridge, Mass.: Harvard University Press.
——— (1986b) Science in Action: How to follow scientists and engineers through society. Milton Keynes: Open University Press.
LAW, J. and WILLIAMS, R. J. (1981) "Social structure and laboratory practice." Paper read at conference on Communication in Science, Simon Fraser University, 1-2 September.
——— (1982) "Putting facts together: a study of scientific persuasion." Social Studies of Science 12: 535-557.
LEPENIES, W. (1981) "Anthropological perspectives in the sociology of science." Pp. 245-261 in Mendelsohn and Elkana (1981).
* LIN, K. C., LIESHOUT, P. v., MOL, A., PEKELHARING, P. and RADDER, H. (1982) Krisis: Tijdschrift voor Filosofie 8: 88-96.
* LONG, D (1980) American Scientist 68: 583-584.
LYNCH, MICHAEL E. (1982) "Technical work and critical inquiry: investigations in scientific laboratory." Social Studies of Science 12: 499-533.

LYNCH, MICHAEL (1985a) "Discipline and the material form of images: an analysis of scientific visibility." Social Studies of Science 15: 37-66.

—— (1985b) Art and Artifact in Laboratory Science: A Study of Shop Work an Shop Talk in a Research Laboratory. London: Routledge and Kegan Paul.

McKEGNEY, DOUG (1982) Local Action and Public Discourse in Animal Ecology: a communcations analysis of scientific inquiry. M.A. dissertation, Simon Fraser University.

MENDELSOHN, E. and ELKANA, Y. [eds.] (1981) Sciences and Cultures. Sociology of the Sciences Yearbook, Vol. 5. Dordrecht and Boston, Mass.: Reidel.

MULKAY, MICHAEL (1984) "The scientist talks back: a one-act play, with a moral, about replication in science and reflexivity in sociology." Social Studies of Science 14: 265-282.

—— POTTER, JONATHAN and YEARLEY, STEVEN (1983) "Why an analysis of scientific discourse is needed." Pp. 171-203 in Knorr-Cetina and Mulkay (1983).

* MULLINS, N. (1980) Science Technology and Human Values 30: 5.

NICKLES, THOMAS (1982) Review of Karin Knorr-Cetina's The Manufacture of Knowledge. Society for Social Studies of Science Newsletter 7 (4): 35-39.

—— (1984) "A revolution that failed: Collins and Pinch on the paranormal." Social Studies of Science 14: 297-308.

PERRY, STEWART E. (1966) The Human Nature of Science: Researchers at Work in Psychiatry. New York: Macmillan.

SCHUTZ, ALFRED (1944) "The Stranger."American Journal of Sociology 50: 363-376. Reprinted pp. 91-105 in Schutz, Collected Papers II: Studies in Social Theory, ed. Arvid Brodersen (1964). The Hague: Martinus Nijhoff.

STAR, SUSAN LEIGH (1983) "Simplification in scientific work: an example from neuroscientific research." Social Studies of Science 13: 205-228.

* STEWART, J. (1982) "Facts as commodities?" Radical Science Journal 12: 129-140.

* THEMAAT, V. V. (1982) Zeitschrift für allgemeine Wissenschafstheorie 13: 166-170.

TIBBETTS, PAUL and JOHNSON, PATRICIA (forthcoming) "The discourse and praxis models in recent reconstructions of scientific knowledge generation." Social Studies of Science.

* TILLEY, N. (1981) "The logic of laboratory life." Sociology 15: 117-126.

TRAWEEK, SHARON (1980) "Culture and the organisation of scientific research in Japan and the United States." Journal of Asian Affairs 5: 135-148.

—— (1981) "An anthropological study of the construction of time in the high energy physics community." Paper, Program in Science, Technology and Society, Massachusetts Institute of Technology.

—— (forthcoming) Particle Physics Culture: Buying Time and Taking Space.

WADE, NICHOLAS (1981) The Nobel Duel. New York: Doubleday.

WILLIAMS, R. J. and LAW J. (1980) "Beyond the bounds of credibility." Fundamenta Scientiae 1: 295-315.

* WESTRUM, R. (1982) Knowledge 3 (3): 437-439.

* WOLFF, R. D. (1981) "Science, empiricism and marxism: Latour and Woolgar vs. E. P. Thompson." Social Text 4: 110-114.

WOOLGAR, STEVE (1981a) "Science as practical reasoning." Paper read at conference on Epistemologically Relevant Internalist Studies of Science, Maxwell School, Syracuse University, 10-17 June.

—— (1981b) "Documents and researcher interaction: some ways of making out what is happening in experimental science." Paper read at conference on Communication in Science, Simon Fraser University, 1-2 September.

WOOLGAR, STEVE (1982) "Laboratory Studies: a comment on the state of the art." Social Studies of Science 12: 481-498.

——— (1983) "Irony in the social study of science." Pp. 239-266 in Knorr-Cetina and Mulkay (1983).

——— (1984) "A kind of reflexivity." Paper read to Discourse and Reflexivity Workshop, University of Surrey, 13-14 September. Forthcoming in Cultural Anthropology.

——— (forthcoming) Science As Practical Reasoning: the practical management of epistemological horror.

ZENZEN, M. and RESTIVO, S. (1982) "The mysterious morphology of immiscible liquids: a study of scientific practice." Social Science Information 21: 447-473.

INDEX